42 Topics in Current Chemistry
Fortschritte der chemischen Forschung

New Concepts II

 Springer-Verlag Berlin Heidelberg GmbH 1973

This series presents critical reviews of the present position and future trends in modern chemical research. It is addressed to all research and industrial chemists who wish to keep abreast of advances in their subject.

As a rule, contributions are specially commissioned. The editors and publishers will, however, always be pleased to receive suggestions and supplementary information. Papers are accepted for "Topics in Current Chemistry" in either German or English.

Any volume of the series may be purchased separately.

ISBN 978-3-662-15990-3   ISBN 978-3-540-37729-0 (eBook)
DOI 10.1007/978-3-540-37729-0

Originally published by Springer-Verlag Berlin Heidelberg New York in 1973.
Softcover reprint of the hardcover 1st edition 1973
Library of Congress Catalog Card Number 51-5497.
Typesetting and printing: Hans Meister KG, Kassel.

Contents

Qualitative and Semiquantitative Evaluation of Reaction Paths
M. Simonetta   1

Graph Theory and Molecular Orbitals
I. Gutman and N. Trinajstić 49

The Electrostatic Molecular Potential as a Tool for the Interpretation of Molecular Properties
E. Scrocco and J. Tomasi 95

Qualitative and Semiquantitative Evaluation of Reaction Paths

Prof. Massimo Simonetta

Istituto di Chimica Fisica dell'Università e Centro del C.N.R., Milano, Italy

Contents

I.   Introduction   2

II.   Orbital Symmetry Rules   3

III. Semiquantitative Calculations   24

IV. Summary and Conclusions   41

V.   Bibliography   42

I. Introduction

Until a few years ago theoretical chemistry could have been more properly called theoretical molecular physics. Most papers in the field had the object of calculating in a more or less sophisticated way the properties of the ground state of simple, isolated molecules, or the difference in energy and transition probability between the ground state and low-lying excited states.

Frequently the work involved conjugated molecules to which σ—π approximation could conveniently be applied, thus drastically reducing the number of electrons in the calculation. Electronic population analysis was usually added to the energy calculations and a theoretical dipole moment was obtained that could be compared with the experimental data. With the advent of NMR. and ESR. spectroscopy other observables became available, and theory was successfully applied to the interpretation of these spectra. However, very little was done in the field of real chemistry, that is, in the study of reaction mechanisms and reaction rates. Over the last decade the availability of large electronic computers, the introduction of approximate but reliable quantum mechanical methods which include all the electrons, or at least all valence electrons in a molecular system and the discovery of the rules of orbital symmetry have led to a significant change of the situation.

With these tools at hand, theoretical chemists were in a position to consider the reactivity problem in a new perspective. It became possible to predict which reaction mechanisms would lead to reasonable or to abnormal energy barriers, and to calculate with acceptable accuracy barrier heights, complete potential surfaces, or even trajectories on these surfaces. The geometry and electronic structure of short-lived species such as activated complexes or unstable reaction intermediates were obtained this way. This gave the experimental scientist invaluable information about species not amenable to experimental investigations.

The aim of the present review is to provide chemists with a general survey of the different techniques now available for the theoretical evaluation of reaction paths. Qualitative work is based nowadays mainly on orbital symmetry rules; this topic is given special emphasis here, since the method is of general use in everyday chemistry. Methods that require actual computation are described in the second part of this review under the heading semi-quantitative methods, since a complete, non-approximate, quantum-mechanical calculation of a reaction rate has never yet been carried out, even for the simplest systems.

II. Orbital Symmetry Rules [1]

In 1965 R.B. Woodward and Roald Hoffmann [2] defined electrocyclic reactions as the two following processes:

In reaction I → II a linear system containing $k\,\pi$ electrons closes to a ring system containing $k-2\,\pi$ electrons, while the number of σ bonds is increased by one. Reaction I → II can occur following two different paths, a "disrotatory" one, III → IV

and a "conrotatory" one: V → VI

Among other examples, they quote the thermal isomerization of cyclobutenes as conrotatory, and the cyclization of hexatrienes to cyclohexadienes, which is disrotatory if obtained thermally and conrotatory when obtained photochemically.

The explanation offered for the stereospecificity of electrocyclic transformations was in terms of the properties of the highest occupied Hückel molecular orbital (HOMO) of the open chain molecule.

3

For thermal reactions, the HOMO is the bonding orbital of highest energy. In systems containing $4n$ electrons the symmetry of this orbital is such that positive overlap between the π orbitals centered on the two terminal atoms obtains through the conrotatory motion, while in systems containing $4n + 2$ electrons the favourable motion is disrotatory, as can

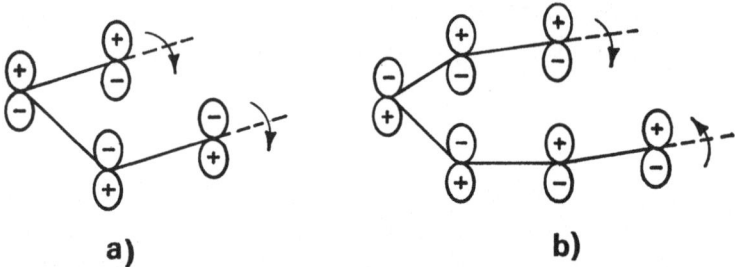

a)   b)

Fig. 1. a) Conrotatory motion in $4n$ electron systems; b) disrotatory motion in $4n + 2$ electron systems

be seen in Fig. 1. On the other hand, excitation of one electron to the lowest unoccupied molecular orbital (LUMO) reverses the symmetry relationship at the two terminal atomic orbitals; thus reactions that proceed thermally by a conrotatory path follow a conrotatory course when photochemically excited ,and vice versa.

Of course, the symmetry rule specifies which of the two possible courses is favored without excluding the other one; the alternative can be favored by other factors which the simple HMO theory does not take into account. For example, cis 1,2,3,4-tetramethylcyclobut-1-ene conrotatorily transforms into cis-trans tetramethylbutadiene when heated at 200 °C. But in dimethylbicyclo[3.2.0]-heptene the presence of the five-membered ring makes the conrotatory process impossible, and the opening reaction to 3-dimethyl-cycloheptadiene occurs at 400 °C, probably via a disrotatory process.

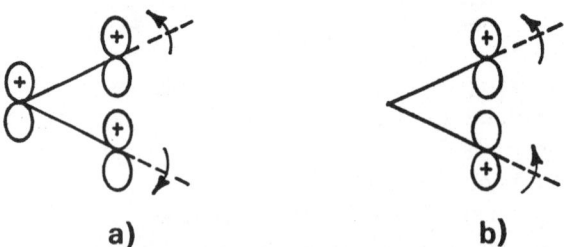

a)   b)

Fig. 2. a) HOMO in the allyl cation; b) HOMO in the allyl radical and anion

4

By these simple rules Woodward and Hoffmann predicted a disrotatory course for the opening of the cyclopropyl cation in its ground state to the corresponding allyl cation, while the thermal opening of cyclopropyl radical and anion to allyl radical and anion is conrotatory. A glance at Fig. 2 clearly shows the reason. Reverse predictions can be made for photochemically induced reactions.

Note that for every electrocyclic reaction there are two conrotatory and two disrotatory motions that may or may not be distinguishable. For example, the two conrotatory motions for *trans* 3,4-dimethylcyclobut-1-ene lead to *cis-cis* and *trans-trans*-1,4-dimethylbutadiene:

In fact, only the *trans-trans* isomer is obtained, probably because of steric factors.

A few months after the communication by Woodward and Hoffmann, H. C. Longuet-Higgins and E. W. Abrahamson [3] suggested a different approach to the problem of stereospecificity of electrocyclic reactions, based on correlation diagrams between the orbitals, the electron configurations, and the states of the reactant and the product.

Let us consider again the cyclobutene-butadiene reaction. The molecular orbitals of cyclobutene that undergo a radical change in the course of reaction are σ and σ^*, the bonding and antibonding orbitals of the bond that is broken, and π and π^* the bonding and antibonding orbitals of the double bond.

The corresponding orbitals in butadiene are $\psi_1 \psi_2 \psi_3 \psi_4$, that is, the two π bonding and the two π^* antibonding orbitals, in order of increasing energy. The two sets of orbitals can be correlated on the basis of their symmetry. Consider the reaction along the conrotatory path. The system preserves throughout the reaction a twofold axis of symmetry C_2, and each orbital may be classified as A or B, that is, symmetric or antisymmetric about this axis (Fig. 3). In the disrotatory reaction

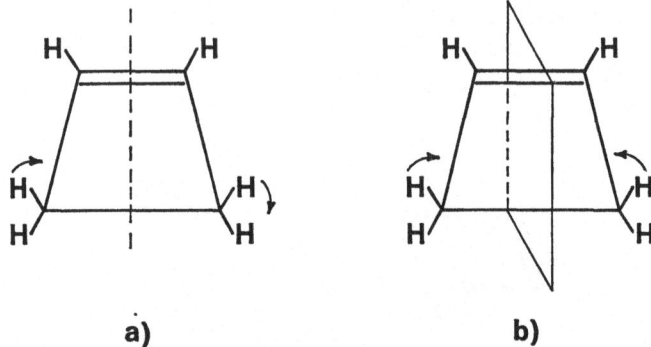

a)   b)

Fig. 3. The C_2 axis and the symmetry plane in cyclobutane. The arrows indicate the motion of atoms in the conrotatory (a) and disrotatory (b) modes

a symmetry plane is mantained (Fig. 3) and the orbitals may be classified as symmetric (A') or antisymmetric (A") with respect to this plane. The orbitals of butadiene can be classified in the same way and the results for reactant and product are shown in Table 1 and Fig. 4.

Table 1. Classification of orbitals in cyclobutene and *cis* butadiene

	Symmetry	cyclobutene	*cis* butadiene
Conrotatory path	A	σ, π^*	ψ_2, ψ_4
	B	π, σ^*	ψ_1, ψ_3
Disrotatory path	A'	σ, π	ψ_1, ψ_3
	A"	π^*, σ^*	ψ_2, ψ_4

From the correlation between orbitals we can derive the correlation between electron configurations. The results are given in Fig. 5. The broken arrows correlate electron configurations. However, electron repulsion prevents states of the same symmetry from crossing, and the full arrows represent correlations between the states. It appears that the

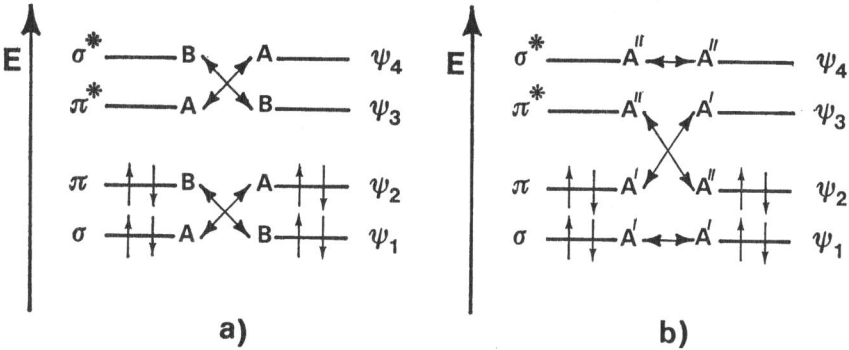

Fig. 4a and b Correlation diagram for orbitals in the cyclobutene–butadiene inter-conversion; a) conrotatory motion; b) disrotatory motion

reaction should follow the conrotatory path for molecules in the ground state and the disrotatory path for molecules in the first excited state.

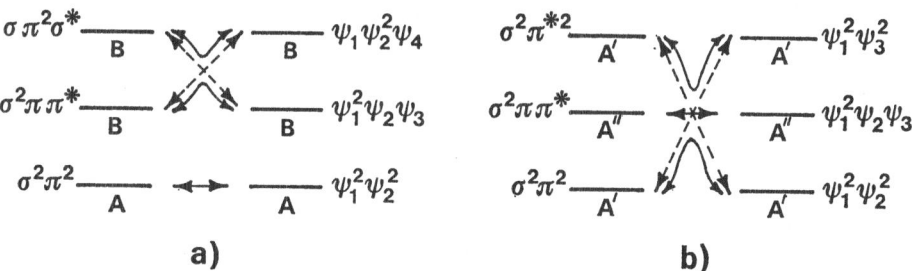

Fig. 5a and b. Correlation diagram for electron configurations and states in the cyclo-butene–butadiene interconversion; a) conrotatory motion, b) disrotatory motion

The same conclusion had been derived by the Woodward-Hoffmann approach. For the hexatriene–cyclohexadiene system too, the results are coincident for the two approaches. The situation is slightly different in cyclopropyl–allyl systems. The two methods give the same results for ions but for the radicals, both in the conrotatory and disrotatory modes, the ground state of each radical is correlated with an excited state of the other. The thermal reactions should prefer the conrotatory to the disrotatory path but would require a larger activation energy than the corresponding transformation in the cations and anions. The conrotatory motion should also be preferred in the photochemical reactions. The correlation diagrams for the orbitals and the electron configurations and states are shown in Figs. 6 and 7.

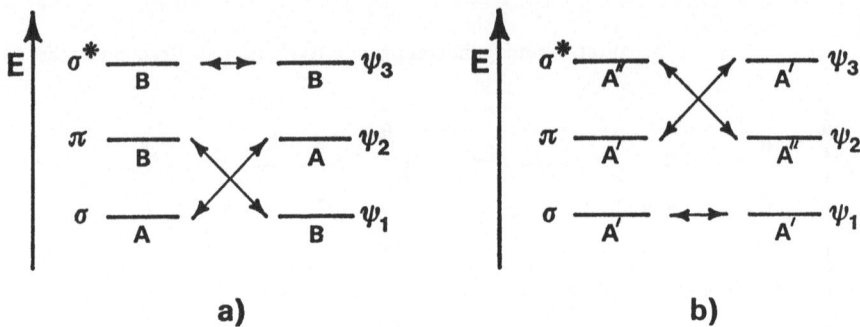

Fig. 6a and b. Correlation diagram for the orbitals in the cyclopropyl–allyl systems; a) conrotatory motion; b) disrotatory motion

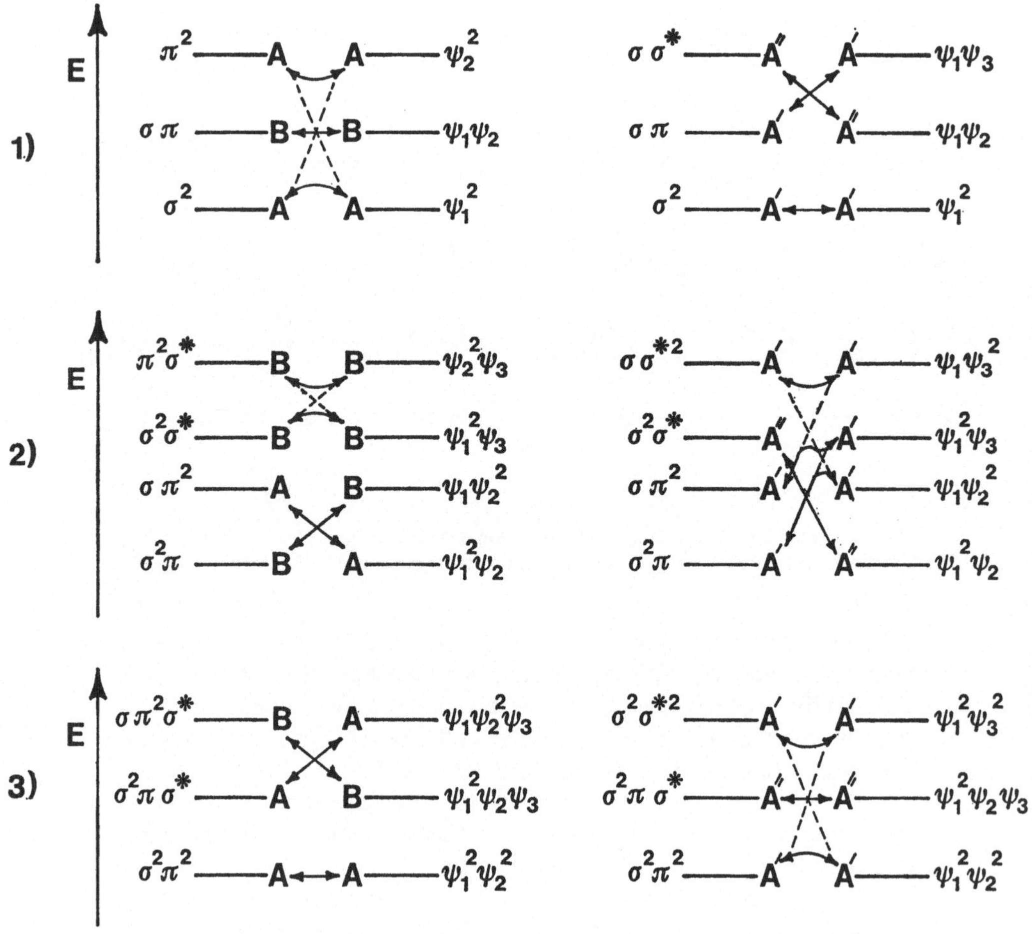

Fig. 7a and b. Interconversion of cyclopropyl and allyl cation (1), radical (2), and anion (3): correlation diagrams for electron configurations and states; a) conrotatory motion; b) disrotatory motion

Woodward and Hoffmann had been privately informed of the method of correlation diagrams [3] and published in the same issue of the same journal a study of concerted intermolecular cycloaddition reactions, for which selection rules were established by means of correlation diagrams for molecular orbitals [4].

First, let us consider the case of the formation of cyclobutane from two ethylenes: the two reacting molecules lie one above the other in parallel planes (Fig. 8). The molecular orbitals may be classified with

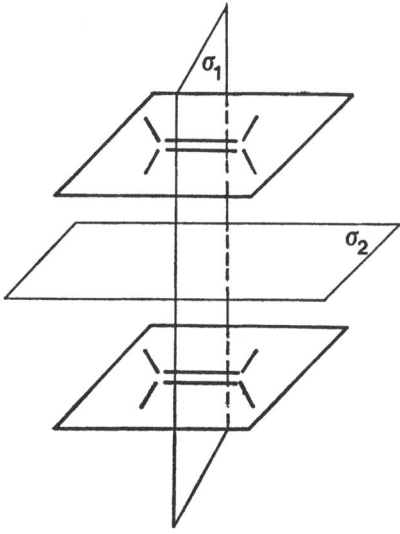

Fig. 8. The approach of two ethylene molecules to form cyclobutane. The symmetry planes σ_1 and σ_2 are also shown

respect to two planes of symmetry, σ_1, a plane bisecting the ethylenes, and σ_2, a plane parallel to the planes of the ethylenes and midway between them. In the reactants we have two orbitals of type π and two orbitals of type π^*. In the product we have instead two bonding orbitals σ and two antibonding orbitals σ^*. To make the symmetry classification possible, we have to consider the combinations $\pi_1 \pm \pi_2$, $\pi_1^* \pm \pi_2^*$, $\sigma_1 \pm \sigma_2$, $\sigma_1^* \pm \sigma_2^*$. The projection of these orbitals upon the plane containing the four carbon atoms, orthogonal both to σ_1 and σ_2, is shown together with the symmetry classification with respect to σ_1 and σ_2 in Fig. 9. The correlation diagram for these orbitals is shown in Fig. 10. The orbitals π SS and π^* AS, which are bonding in the region of the reaction, decrease in energy along the reaction coordinate, while the orbitals π SA and π^*

M. Simonetta

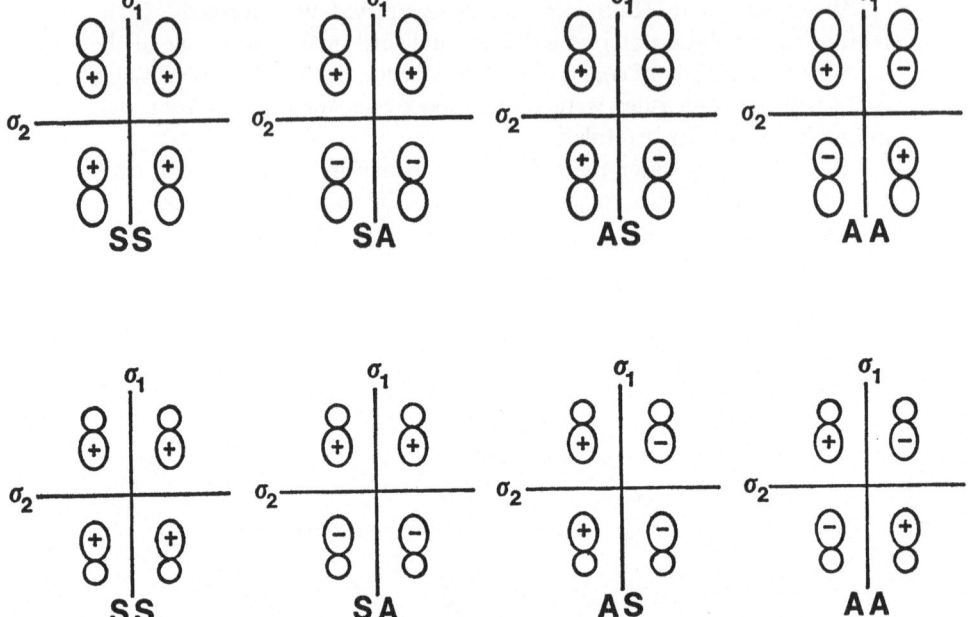

Fig. 9. Classification of orbitals in cyclobutane and two ethylenes

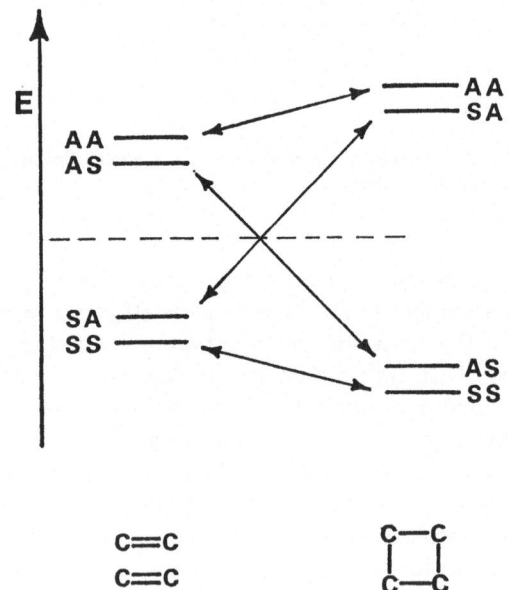

Fig. 10. Correlation diagram for molecular orbitals for two ethylenes and cyclobutane

10

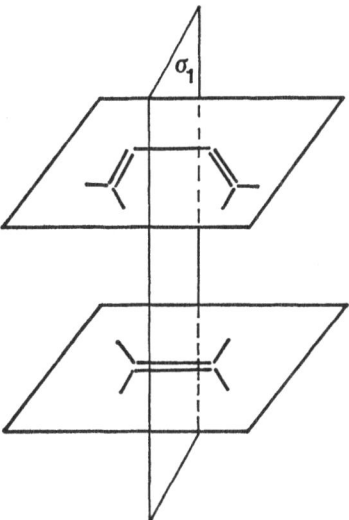

Fig. 11. The approach of ethylene and butadiene

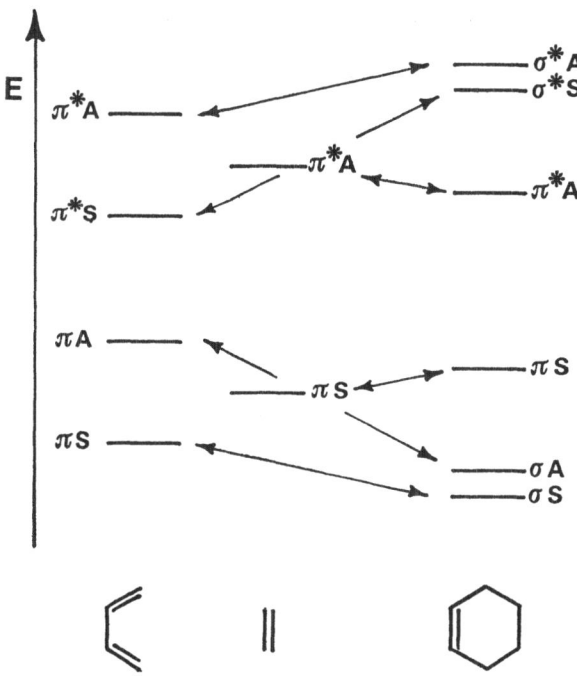

Fig. 12. Correlation diagram for the reaction ethylene + butadiene → cyclohexene

AA that are antibonding in the same region increase in energy. We may notice a crossing of bonding and antibonding levels. Let us consider now the addition of ethylene to butadiene. The two molecules approach lying in parallel planes, and there is now only one plane of symmetry, σ_1 (Fig. 11). The levels are classified as symmetric or antisymmetric with respect to σ_1. The correlation diagram is reported in Fig. 12. The three orbitals of S symmetry interact: the one with the lowest energy decreases in energy, the highest one increases, and the middle one remains almost constant. A similar interaction occurs with the three A orbitals, but there is no crossing of bonding and antibonding levels here.

If we compare Figs. 10 and 12, we see that 1,2 additions are symmetry-forbidden processes in the ground state, while 1,4 additions are symmetry-allowed. If one electron is excited to the next higher level, the energy along the reaction path is decreased for 1,2 additions and increased for 1,4 additions. A reasonable generalization leads to the statement that correlation diagrams without bonding–antibonding crossing are characteristic of allowed thermal reactions, while diagrams in which bonding levels are correlated with antibonding levels are typical of photochemical processes.

The following rules may be derived, in which m, n, p, are numbers of π electrons, and q is an integer, null or positive:

a) electrocyclic reactions, in which one π bond disappears and one σ bond forms, $1\pi \rightarrow 1\sigma$

σ or C_2

The allowed reactions are disrotatory if thermal, or conrotatory if photochemical when m, the number of π electrons in the open molecule is $4q+2$, and vice versa when $m=4q$.

b) reactions of type $2\pi \rightarrow 2\sigma$ are thermally allowed if $m+n=4q+2$, and photochemically allowed when $m+n=4q$.

c) reactions of type $3\pi \rightarrow 3\sigma$, with a plane of symmetry bisecting m are thermally allowed if $m = 4q + 2$ (any p) and photochemically allowed when $m = 4q$ (any p).

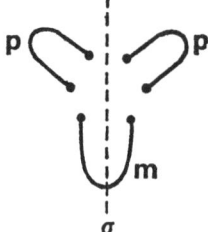

d) reactions of type $4\pi \rightarrow 4\sigma$ are thermally allowed for $m + n = 4q + 2$ (any p), photochemically for $m + n = 4q$ (any p).

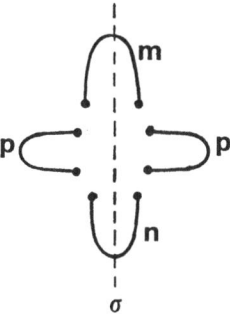

These rules can easily be checked by counting the numbers of levels of a given symmetry (*e.g.* S) in reactants and products. If the numbers are equal the reaction is thermally allowed, since there are no bonding–antibonding crossings. Let us consider, *e.g.* $2\pi \rightarrow 2\sigma$ reactions. In a reagent molecule with m π orbitals there are $m/4$ bonding π orbitals with S symmetry if $m/2$ is even (*e.g.* butadiene, $m = 4$, $m/2 = 2$,

$m/4 = 1$), but $\dfrac{m+2}{4}$ such orbitals if $m/2$ is odd (*e.g.* ethylene, $m = 2$, $m/2 = 1$, $\dfrac{m+2}{4} = 1$; hexatriene, $m = 6$, $m/2 = 3$, $\dfrac{m+2}{4} = 2$). The same component in the product will contribute $m - 2\,\pi$ orbitals, and so $m/4$ bonding π orbitals of S symmetry if $m/2$ is even, and hence $\dfrac{m-2}{2}$ is odd, but $\dfrac{m-2}{4}$ bonding π orbitals of S symmetry if $m/2$ is odd, and hence $\dfrac{m-2}{2}$ is even. The σ levels in the product will contribute one bonding level of S symmetry. It follows that in thermally allowed reactions the number of bonding π orbitals of S symmetry in the reagent must be greater by one than the number of the orbitals of the same kind in the products.

There are three possibilities:

1) if $m = 4q_1$ and $n = 4q_2$, the number of bonding π orbitals of S symmetry is $q_1 + q_2$ before and after reaction

2) if $m = 4q_1 + 2$, $n = 4q_2$ (or vice versa), we have $q_1 + q_2 + 1$ levels before and $q_1 + q_2$ levels after the reaction

3) if $m = 4q_1 + 2$, $n = 4q_2 + 2$, the numbers of levels are $q_1 + q_2 + 2$ and $q_1 + q_2$.

It is immediately seen that only in case 2 is the reaction thermally allowed. The condition can be reformulated as follows:

$$n + m = 4q_1 + 4q_2 + 2 = 4q + 2 \text{ where } q = q_1 + q_2 .$$

The smaller allowed, concerted cycloaddition reactions are collected in Table 2.

Table 2. Allowed cycloaddition reactions

Type of reaction	Thermally allowed				Photochemically allowed			
	m	n			m	n		
$2\,\pi \to 2\,\sigma$	4	2			2	2		
	6	4			4	4		
	8	2			6	2		
	m	p	p		m	p	p	
$3\,\pi \to 3\,\sigma$	2	2	2		4	2	2	
	2	4	4					
	6	2	2					
	m	n	p	p	m	n	p	p
$4\,\pi \to 4\,\sigma$	4	2	2	2	2	2	2	2

The rules hold for all concerted cycloadditions, even when the rates at which the new σ bonds form differ considerably. However, the rules cannot be expected to hold in cases where a multistep reaction occurs, with formation of biradical or dipolar intermediates.

In cycloadditions considered up to now, in each reagent the bonds are formed or broken on the same side of the molecule:

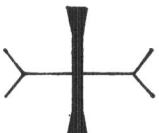

The reactions are then said to take place in a suprafacial manner [5] and are examples of $_\pi 2_s + _\pi 2_s$ and $_\pi 4_s + _\pi 2_s$ processes.

However, there are alternative processes, called antarafacial processes, in which bonds are formed or broken on opposite faces of the reagent molecule. Let us consider again the addition of two ethylene molecules to form cyclobutane but with a different geometrical approach:

If we combine the π orbital of one molecule with a π^* orbital of the other molecule and vice versa, cyclobutane may be formed by the following process:

This is called a $_\pi 2_s + _\pi 2_a$ process. We can classify the orbitals of the reactants and of the product with respect to a C_2 axis perpendicular to the C—C bonds in both ethylene molecules. If molecule 1 is the underlying ethylene the classification is as follows: π_1 and π_2^* are symmetric,

15

π_1^* and π_2 antisymmetric. In the same way we can classify the bonds in cyclobutane: $\sigma_1 + \sigma_2$ and $\sigma_1^* + \sigma_2^*$ are symmetric, $\sigma_1 - \sigma_2$ and $\sigma_1^* - \sigma_2^*$ antisymmetric. The correlation diagram shown in Fig. 13 is obtained. It is evident that while the $_\pi 2_s + _\pi 2_s$ process is thermally forbidden, the $_\pi 2_s + _\pi 2_a$ process is thermally allowed. We have seen that the $4_s + 2_s$

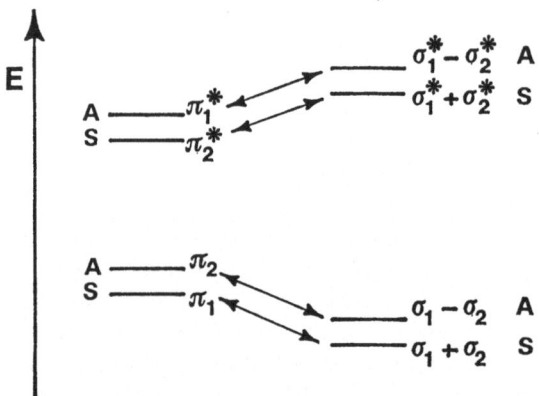

Fig. 13. Correlation diagram for the antarafacial process ethylene + ethylene → cyclobutane

process is thermally allowed. This process can also be described as a $2_s + 2_s + 2_s$ process.

The following rule can be derived for concerted cyclo-addition reactions. They can be described as $2 + 2 + 2 + \ldots$ processes. Then we may count the number of suprafacial and antarafacial processes. If the number of suprafacial processes is odd, the reaction is symmetry-allowed; if it is even, the reaction is symmetry-forbidden. It must be borne in mind, however, that antarafacial processes are frequently hindered by steric factors.

Orbital symmetry relationships can be useful in the study of secondary conformational effects in concerted cyclo-addition reactions [6]. One example is the Diels-Alder addition of butadiene to itself. This $_\pi 4_s + _\pi 2_s$ reaction may take place through an endo route (a) or an exo route (b), differing mainly by the proximity of a β and a β' orbital in the endo approach. The secondary interaction among occupied orbitals in the two reactions will be negligible and the significant interaction will come from symmetry-allowed mixing of occupied orbitals in one reactant with unoccupied orbitals of the other. It is easily seen that mixing of both

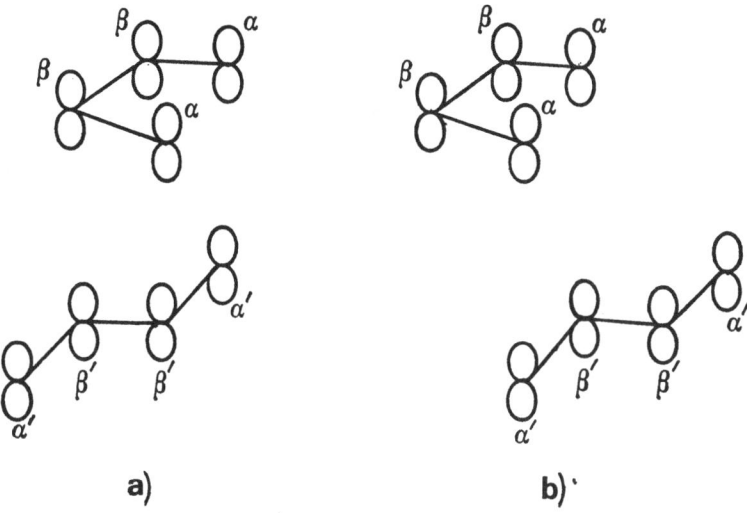

the HOMO of diene with LUMO of olefin and the HOMO of olefin with LUMO of diene leads to a bonding interaction of the β and β' orbitals close to each other (Fig. 14).

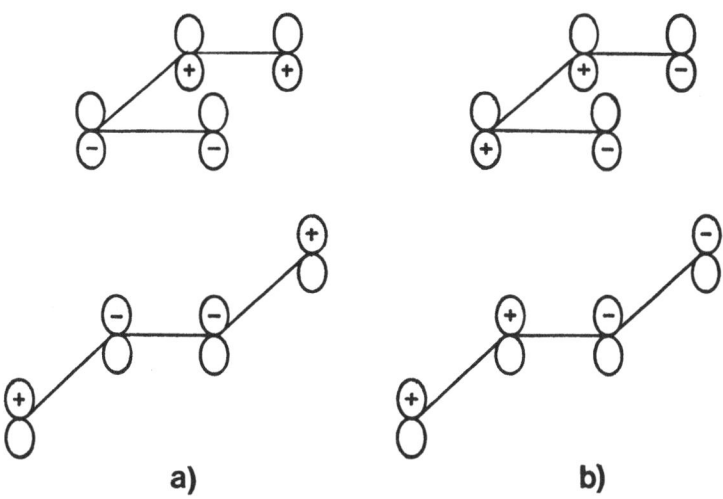

Fig. 14a and b. Mixing of orbitals in the butadiene + butadiene Diels-Alder reaction

The endo transition state is favored. Fig. 15 shows that the opposite is true in the case of the symmetry allowed $_\pi 6_s + _\pi 4_s$ additions.

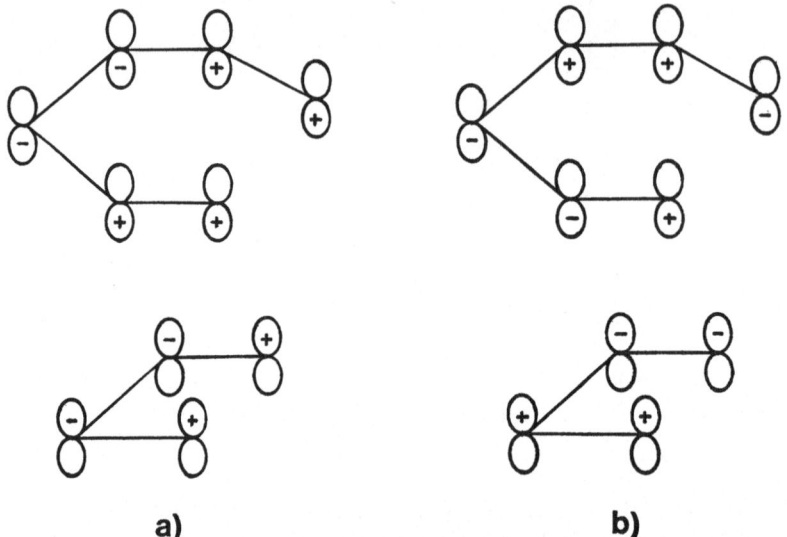

a)   b)

Fig. 15a and b. Mixing of orbitals in the $_\pi 6_s + _\pi 4_s$ reation

Now we consider sigmatropic reactions. Woodward and Hoffmann defined a sigmatropic reaction of order $[i,j]$ as the migration of a σ bond flanked by one or more π electron systems, to a new position removed by $i-1$ and $j-1$ atoms from the original bonded sites in an uncatalyzed intramolecular process [7].

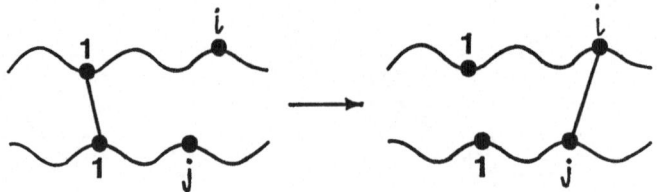

An example of a [3,3] sigmatropic reaction is the Cope rearrangement:

$$CH_2-CH=CH_2 \qquad CH_2=CH-CH_2$$
$$| \qquad\qquad\qquad\longrightarrow \qquad |$$
$$CH_2-CH=CH_2 \qquad CH_2=CH-CH_2$$

We start by considering the [1,j] sigmatropic migration of a hydrogen atom within an all-*cis* polyolefin:

$$\overset{1\ \ 2}{R_2C}{=}CH{-}(CH{=}CH)_k{-}\overset{j}{C}HR_2' \rightarrow \overset{1}{R_2CH}{-}(CH{=}CH){-}_kCH{=}\overset{j}{C}R_2'$$

In the transition state we will have a hydrogen atom loosely bonded to a radical containing $2k+3\ \pi$ electrons. The process may be suprafacial, with the hydrogen atom always on the same face of the π system and the transition state showing a plane of symmetry, or antarafacial, with the hydrogen atom bond first to one face and then to the other face of the π system, the transition state showing a C_2 axis of symmetry, and the hydrogen atom in the plane of the radical carbon skeleton. The HOMO radical has the following symmetry:

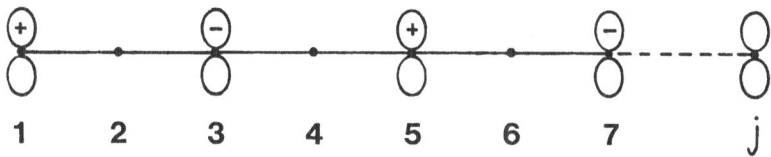

Since the hydrogen bonding orbital is a 1s orbital, a thermal reaction will follow the suprafacial route if k is odd, and the antarafacial route if k is even; *e.g.*

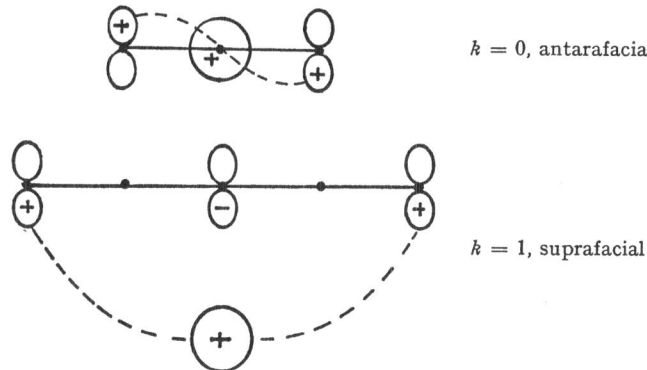

$k = 0$, antarafacial

$k = 1$, suprafacial

The reverse conclusion would be obtained for sigmatropic migrations taking place within species in the first excited state. The symmetry-allowed [1,j] sigmatropic reactions for $j \leqslant 7$ are collected in Table 3.

Table 3. Symmetry allowed [1,j] sigmatropic reactions

[1, j]	Thermal Reaction	Photochemical Reaction
[1, 3]	Antarafacial	Suprafacial
[1, 5]	Suprafacial	Antarafacial
[1, 7]	Antarafacial	Suprafacial

In the foregoing it has been assumed that the migrating group interacts with the π system by means of a σ orbital (an s orbital in the case of hydrogen). If the migrating group interacts with the π system by means of a π orbital, relationships reversed from those of Table 3 can occur:

When both j and $i > 1$ and the transition state has a plane of symmetry, thermal processes are allowed if $i + j = 4n + 2$, and first-excited states processes are allowed if $i + j = 4n$. E.g. a [3,3] process is thermally allowed:

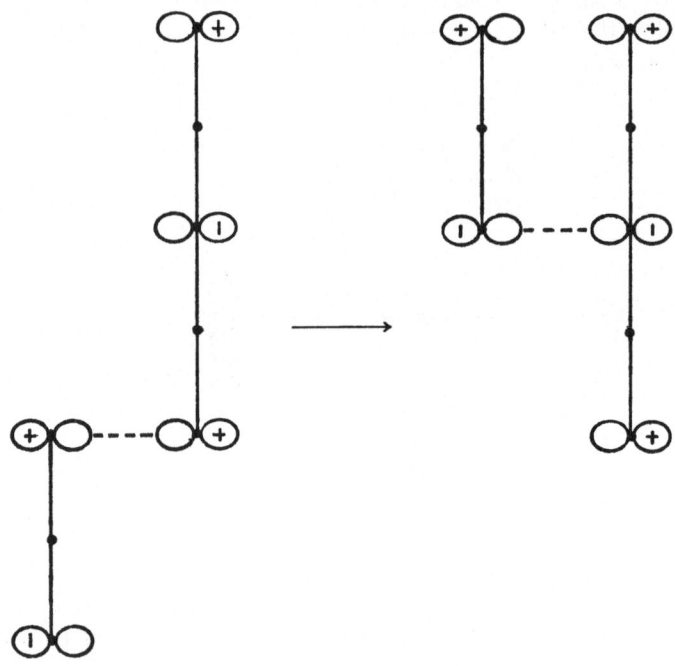

It is of special interest that the orbital symmetry rules can be applied to sigmatropic migrations within ionic species. *E.g.* the suprafacial 1,2 shift within a carbonium ion is symmetry-allowed:

$$\overset{\oplus}{CH_2}-CH=CH_2 \longrightarrow CH_2=\overset{\oplus}{C}-CH_3$$

The symmetry rules predict that 1,3 suprafacial, thermal sigmatropic migration of carbon should proceed with inversion of configuration of the migrating group. A very interesting example of 1,3 sigmatropic reaction is the rearrangement of the ketone exo-3-methylbicyclo[4.2.0] oct-5-en-2-one to give exclusively exo-3-methylbicyclo[2.2.2] oct-5-en-2-one [8].

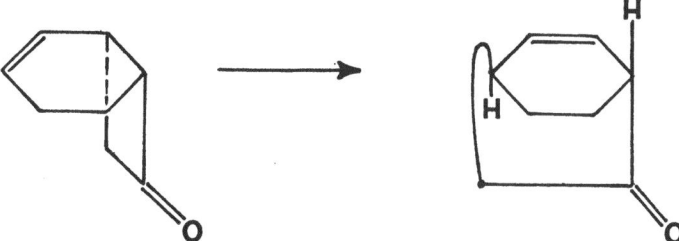

The antarafacial process here can be excluded since it would produce a highly strained molecule (Fig. 16). In such rearrangements for bicyclo-heptenes and bicyclooctenes there are other examples, however, where the reaction is not so highly stereospecific, or even where retention is favoured over inversion. In such cases one might invoke a nonconcerted

Fig. 16. The product of the antarafacial process for [1,3]sigmatropic migration in bicyclo[4.2.0]oct-2-en-7-one

mechanism involving a diradical intermediate. However, simple Hückel-type calculations show that the transition state for the forbidden supra-facial 1,3 sigmatropic rearrangement of carbon, that is, with configuration retention, though less stable than the transition state for the allowed process, is stabilized with respect to the transition state for the diradical

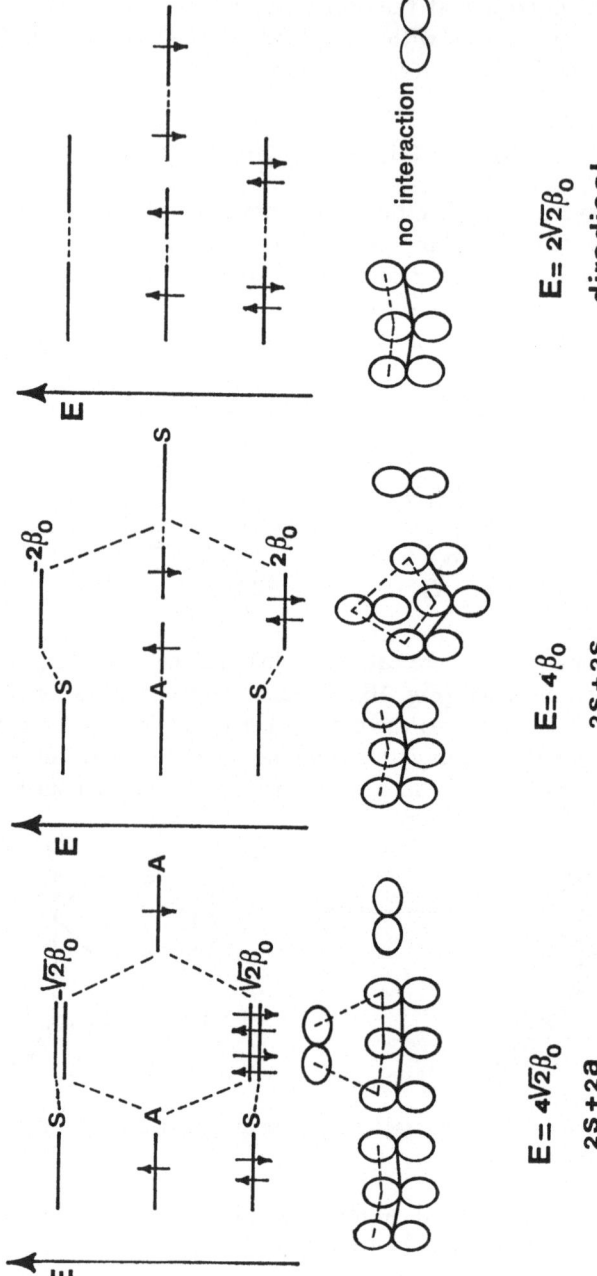

Fig. 17. Energy diagram for different transition states in suprafacial [1,3]carbon migration

mechanism where the p orbital of the migrating carbon does not interact with the allyl radical [9–11]. A diagram of the energy levels obtained with the simplifying assumption of equal nearest-neighbor interaction is shown in Fig. 17. The stabilization of the forbidden transition state comes from the interaction of the carbon p orbital not with HOMO but with a subjacent bonding allyl orbital. Experimental evidence favors the possibility that some forbidden sigmatropic reactions may be concerted [9]. Configuration interaction can also contribute to the stabilization of the transition energy in a forbidden reaction and may offer an explanation for for the occurrence of symmetry-forbidden concerted reactions [12].

The theory of orbital symmetry conservation reported here closely follows the presentation given by Woodward and Hoffmann [1], with the inclusion of the outstanding contribution of Longuet-Higgins and Abrahamson. However, it is only fair to recall that the first suggestion of the role of orbital symmetry in determining the course of stereochemical reactions was put forward by Oosterhoff [13].

The symmetry rules were also extended to the discussion of sigmatropic migrations in C_nH_{n+1} monocycles [14], dielectrocyclic reactions, $i.e.$ reactions in which electrons located in two π orbitals transform by a rotation into electrons in two σ bonds (Fig. 18) [15], reactions of ground state and singlet oxygen with olefins and dienes [16], isomerization and substitution reactions of transition metal complexes [17], sigmatropic rearrangement in organometallic compounds [18]. Among the innumerable

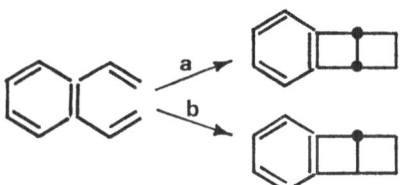

Fig. 18. An example of dielectrocyclic reaction: a) disrotatiory, b) conrotatory

applications, I would like to mention a discussion of several reaction paths for bimolecular hydrogen exchange [19].

There have been many treatments published with the aim of providing a sound quantum mechanical basis to the symmetry rules, or to show how they could be derived by different arguments.

In this connection I would like to mention the work of Fukui [20–22], based on the symmetry of frontier orbitals and the application of perturbation theory [23], and the symmetry rules derived by Pearson for unimolecular reactions, but extensible to reactions of any molecularity [24], based on the idea that a reaction is allowed if the symmetry of the bonds that are made is the same as the symmetry of the bonds that are broken. Alternative demonstrations of the Woodward-Hoffmann rules based on perturbational molecular orbital calculations [25] or the generalized valence bond method [26] and the idea of a phase relationship between the orbitals of the reactants and the orbitals of the products have recently been published. A critical review of the various theoretical derivations of the Woodward-Hoffmann rules and of related approaches was given by Dewar [27].

The applications of symmetry rules have been frequently reviewed [28,29], also at an elementary level [30,31].

Other qualitative rules for the study of reaction paths have been derived independently. For unimolecular reactions, it has been found that conditions favorable to a given path exist if there is a low-energy excited state of the same symmetry as the normal mode corresponding to the reaction coordinate, the transition density is localized in the region of nuclear motion and the excitation energy decreases along the coordinate [32].

These rules have been used in the case of pyrolysis of cyclobutane and of the Diels-Alder retrogression of cyclo-hexene [33]. The same rule has been related to a second order Jahn-Teller effect [34].

An approach very closely related to that of Woodward and Hoffmann is the so-called Hückel-Möbius approach [35] based on the rule: $4n+2$ electron systems prefer Hückel geometries and $4n$ electron systems prefer Möbius geometries [36]. When no symmetry exists and there is no cyclic orbital array the allowedness or forbiddenness of a reaction can be determined by following the form of the MO's during the reaction [37]. A detailed quantum mechanical study of the stereochemistry of thermal and photo cyclo-addition reactions has been reported [38], and a quantum mechanical discussion of the applicability of the Woodward-Hoffmann rules can be found in a paper by George and Ross [39].

III. Semiquantitative Calculations

Up to a few years ago chemical reactivity was discussed in term of reactivity indexes. These approaches, although valuable, will not be discussed here, since they have been frequently reviewed in the past[40–44]. Nor will we discuss the perturbation molecular orbital theory for reactants, which has been the subject of extensive reviews [45–47]. Extensions of this method can be found in papers by Klopman [48–50] and Dougherty [51]. I shall now mention some methods which have not yet found wide popularity but seem very promising. I mean the criterion of maximum localization [52]: here minimum energy sections of potential energy hypersurfaces for reactions of polyatomic molecules are obtained as a function of two variables, chosen among the internal coordinates that undergo large changes from reactants to products, or by the related method using the principle of least motion [53].

For reactions that allow the use of $\sigma-\pi$ approximation, semiempirical methods of calculating potential energy surfaces were elaborated by Basilevsky [54,55] and by Salem [56,57]. An interesting feature of the last method, which stresses the importance of interactions between the top occupied orbital of one molecule and the lowest unoccupied orbital of the other, is that, while there is no need to calculate the whole potential surface, it is possible to derive a theoretical pathway for each reaction. Salem has recently put forward a theory of asymmetric induction, enabling the difference in energy between diastereoisomeric transition states and the diastereoisomer ratio to be calculated for an achiral reagent and a model chiral substrate [58].

Another fairly new method, using the electrostatic molecular potential, will not be discussed here since it is the subject of another contribution to this volume [50]. I will now consider methods that have had the widest application in the theoretical study of chemical reactivity, in order of increasing complexity: a) molecular mechanics; b) extended Hückel method; c), d) empirical self-consistent field methods such as CNDO and MINDO; e) the simplest "ab initio" approach; f) the different S.C.F. methods, possibly including configuration interaction; g) valence bond methods, and h) the dynamical approach, including the calculation of trajectories [61].

For each of these methods I shall discuss one example in some detail, briefly mentioning others, to give a broad idea of their applicability. The literature references are not exhaustive, to keep the present paper within a reasonable length.

By molecular mechanics we mean a method by which we calculate the total energy of a molecule in a particular geometry with reference to a hypothetical molecule with no bond-angle or bond-length deformations, no torsional strain and no steric repulsion and with a given number of single and multiple bonds. The energy difference is obtained as the sum of six components:

$$E_T = \Delta E_\sigma + \Delta E_\pi + E_c + E_b + E_t + E_{nb} \tag{1}$$

where ΔE_σ and ΔE_π represent the variation in the σ and π electron energies, E_c, E_b, E_t and E_{nb} represent bond length, bond angle, torsional and nonbonded interaction strains. For the different terms in (1) a number of empirical functions have been suggested; those given by

Lifson [62,63] and by Bartell [64] seem the most successful. An extension of the method to conjugated molecules in the ground and excited states has recently been published [65]. The method has been widely used in conformational analysis of organic molecules [66,67] and proved a useful tool in the prediction of crystal structures [68,69]. As an example of its application to the study of chemical reactivity, we mention some calculations on sixteen bridgehead systems [70]. The problem in applying molecular mechanics to the study of reactivity is the lack of information on the nature of the transition state. However, for the case of solvolysis of chlorides, bromides, tosylates and triflates of bridgehead systems, the corresponding carbonium ions can be taken as model for the transition states, and the corresponding hydrocarbons as model for the reactants. When the hydrocarbon-cation strain energy differences were plotted against the logarithm of the relative solvolysis rates, good correlation was obtained (see *e.g.* Fig. 19). This can be taken as a suggestion that solvent and entropic effects are constants. Other examples are similar calculations for the solvolysis of series of bromides at 25° in 80% ethanol [71], the evaluation of energy and geometry along the reaction path in the Cope rearrangement of hexa-1,5-diene, *cis* 1,2-divinylcyclo-propane

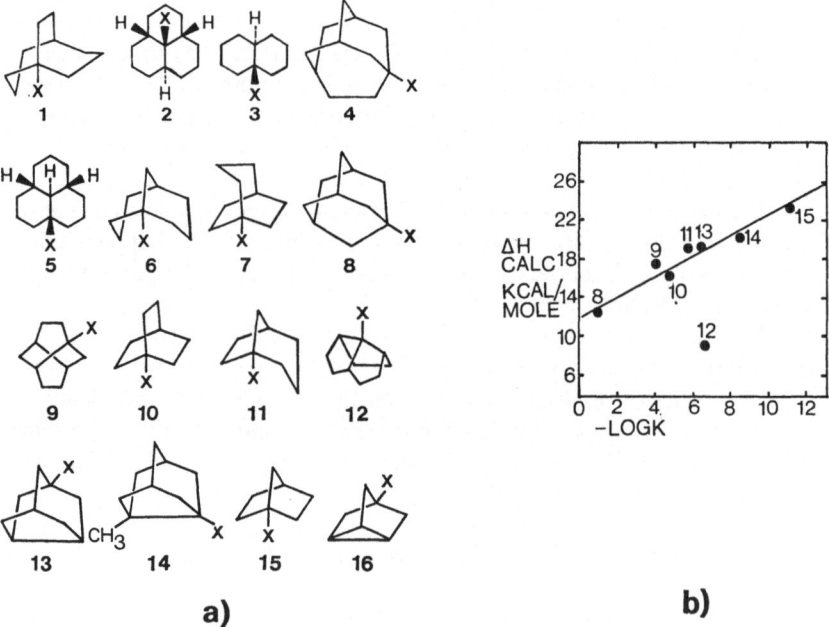

Fig. 19. a) Bridgehead derivatives used in the calculation; b) An example of correlation between -log of the experimental tosylate acetolysis rate constants at 70° and the calculated hydrocarbon–carbonium ion strain energy

and *cis* divinylcyclobutane [72)], and similar calculations for the thermal *cis-trans* isomerization of 2-butene, 2 pentene, β-methylstyrene and stilbene [73)].

The extended Hückel method (EHM) [74)] is the simplest empirical all valence electron molecular-orbital method available at present.

The coefficients c_{ij} of atomic orbitals ϕ_j in molecular orbitals

$$\psi_i = \sum_j c_{ij}\,\phi_j \qquad (2)$$

are obtained by solving the set of equations

$$\sum_i^n |H_{ij} - ES_{ij}|\,c_{ij} = 0. \qquad (3)$$

In hydrocarbons or in molecules containing atoms up to fluorine, only $1s$ hydrogen and $2s$, $2p$ Slater orbitals for the other elements are needed. H_{ii} are chosen equal to valence-state ionization potentials, while H_{ij} are given by

$$H_{ij} = 0.875\,(H_{ii} + H_{jj})\,S_{ij} \qquad (4)$$

where S_{ij} are calculated overlap integrals. Sometimes an exponent 1.3 is used for hydrogen.

While the EHM has known deficiencies as far as quantitative predictions of molecular energies and geometries are concerned, it leads to calculated potential surfaces showing the correct general shape so that qualitative conclusions on the mechanisms of reactions can be safely envisaged.

Among many examples, let us discuss the potential energy surface for the $CH_4 + CH_2$ system [75)]. Two mechanisms have been proposed for the reaction, namely the one-step, or insertion:

$$: CH_2 + R{-}H \longrightarrow R{-}CH_2{-}H$$

and the two-step, or abstraction:

$$: CH_2 + R{-}H \rightarrow .\,CH_3 + .\,R \rightarrow CH_3{-}R.$$

The experimental evidence favors insertion. Calculations were performed for the insertion of singlet methylene, the σ^2 configuration being assumed to be a good approximation of the singlet state.

A symmetry plane was assumed along all the reaction with the transferred hydrogen H_t moving in this plane. For each carbon–carbon distance R, two variables specified the position of H_t, two angles specified the orientation of CH_2 with respect to the CH_3 residual, and one angle specified the flattening of the CH_3 pyramid of methyl. The

C—H bond lengths other than C—H$_t$ were kept constant. For each value of R and each position of H_t the valus of the other three variables were optimized. For $R > 4.0$ Å there are two minima, one corresponding to $CH_4 + CH_2$, and the other, ill-defined, corresponding to $CH_3 + CH_3$. CH_4 is tetrahedral in the first minimum while the two CH_3 radicals are planar. For $R = 3.0$ Å there are still two minima but the barrier between the wells is decreased from 2.5 eV at $R = 4.0$ to 0.8 eV for $R = 3.0$ Å. The well for the $CH_3 + CH_3$ system is now deep and well-defined. Further decrease of R leads to the disappearance of the well for $CH_2 + CH_4$, and we have for $R = 1.5$ Å only one minimum with H_t in the position for the product ethane. Along the reaction path it is found that, for $R > 2.5$ Å, H_t does not move. For $2.4 < R < 2.5$ Å, we have the transfer of H_t. While R changes very little H_t moves from the initial position to one close to the methylene carbon, while the CH_3 group is flattening. The last stage of the reaction is formation of ethane with repyramidalization of the two methyl groups. The transition state corresponds to C—H$_t$—C atoms almost linear; it can be qualitatively represented as

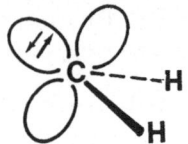

It is not far from abstraction-like attack of methylene on methane. The calculated activation energy is about 7 kcal/mole. The shape of the potential surface also clearly shows that the abstraction-recombination

mechanism is very unlikely. The variations of the bond order and charge distributions are in the expected directions and are consistent with the methylene attacking methane utilizing an empty orbital (Fig. 20).

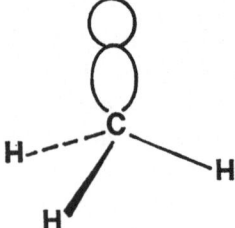

Fig. 20. The approach of methylene to methane along the least-energy path

In conclusion, an illuminating insight of the reaction path is obtained by means of a rather naive, but inexpensive method of computation. Other reactions studied along these lines, include the addition of methylene to ethylene [76], addition and insertion of sulfur to ethylene [77], dimerization of methylenes and nitroso compounds [78] fragmentation of cyclo-butane to ethylenes [79], the Cope rearrangement [80], the interaction of tricycloalkanes with acids and bases [81], and the polytopal rearrangements in $(CH)_5^+$, $(CH)_5^-$, and $(CH)_4CO$ systems [82].

It may be worthwhile mentioning that most of the arguments in the development of the Woodward-Hoffmann rules have been substantiated with the results of EH calculations [1,83]. Other EH molecular orbital studies were devoted to the elucidation of the mechanism of the Wolfe rearrangement [84], and to the investigation of the hydrolysis of acetylcholine [85]. A simplified form of the EH method has been used to determine the role of non-classical ions in 1,2 rearrangements [86].

We now come to discuss methods based on the self-consistent-field theory. After the introduction of the linear combination of atomic orbitals approximation made the application of SCF theory to molecular systems possible [87], it became apparent that, if large molecules were to be treated, other simplifications were needed. When the $\sigma—\pi$ approximation was acceptable, the problem was solved by means of the zero differential overlap approximation [88,89]. However, it was not immediately clear how this approximation could be extended to treatments in which all valence electrons were included. Particular care must be taken to see that the SCF equations are invariant under certain transformations, namely under rotation of local atomic axes and under hybridization of the orbitals on the various atoms. When these transformation properties are recognized, restrictions are imposed on the approximate SCF equations. Two internally consistent approximations were initially proposed: the complete neglect of differential overlap (CNDO), and the neglect of diatomic differential overlap (NDDO) approximations [90]. A number of empirical parameters that appear in the equations can be calibrated according to different criteria. Later other approximations were introduced, such as the partial neglect of differential overlap (PNDO) [91], the intermediate neglect of differential overlap (INDO) [92], and the modified intermediate neglect of differential overlap (MINDO) [94, 95].

The various approximate methods have been extensively described and compared [96–98].

Among these semiempirical SCF methods, the most frequently used in chemical reactivity studies are CNDO/2 (a modified version of CNDO [99], and MINDO/2 [95].

We report here the results of an investigation of the geometry of the CH_5^+ ion by the CNDO/2 method [100]. Considerable interest was aroused by the geometry of this ion, thought to play an important role for the interpretation of the structure of the transition state in electrophilic substitution at alkanes. Three configurations have been explored, which have D_{3h}, C_{4v}, and C_s symmetry (Fig. 21). The following parameters have been

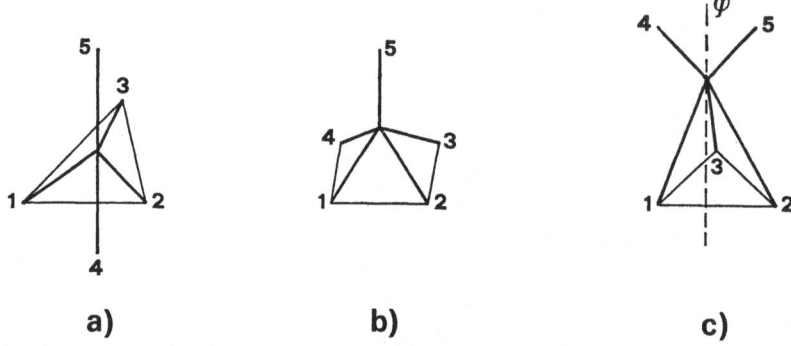

a)   **b)**   **c)**

Fig. 21a-c. Conformations of CH_5^+; a) D_{3h}; b) C_{4v}; c) C_s

allowed to vary: for D_{3h} configuration, C—H_1 and C—H_4 bond lengths; for C_{4v} configuration, the C—H_1, C—H_5 bond lengths and $H_1\widehat{C}H_2$; for configuration C_s, the C—H_1, C—H_4 bond lengths and $H_1\widehat{C}H_2$, $H_4\widehat{C}H_5$. The ternary axis ϕ of the $H_1H_2H_3C$ pyramid and the bisecting line of $H_4\widehat{C}H_5$ were assumed to be coincident. Configuration C_s was found to be the most stable, with the following geometry: $r_{C-H_1} = 1.41$ Å, $r_{C-H_4} = 1.21$ Å, $H_1\widehat{C}H_2 = 110)°$, $H_4\widehat{C}H_5 = 50°$.

If rotation of the H_4—H_5 group around the ϕ axis is allowed, the best geometrical parameters and energy are unchanged. These findings imply that configuration retention is favored on both energetic and entropic grounds. It is worth mentioning here that CH_5^+ and related systems have been the subject of numerous detailed and extensive calculations. The first theoretical studies on the geometrical and electronic structure of CH_5^+ did not consider the possibility of C_s symmetry [101–103]; after the structure with C_s symmetry was introduced [100], a C_{3v} structure was added where the extra proton is attached to one of the CH_4 hydrogens [104]. Besides other CNDO calculations [105,106] in agreement with previous results, a number of "ab initio" calculations appeared [108,109], and the CH_5^- species was also given consideration [110,111]. CNDO studies were extended to the CH_4^+ species [113] and to an examination of the energy surface for CH_5^+ [114] and for the $CH_4 + H$ system [115]. Potential energy surfaces for CH_5^+ and for hydrogen exchange and abstraction in the $CH_4 + H$ system have also been investigated by non-empirical calculations [116–118].

The general conclusions that may be drawn are that CH_5^+ is stable with C_s symmetry, and the approach of H_2 along the C_3 axis of CH_3^+ holds with no activation energy. In the $CH_4 + H$ system at low energy axial abstraction is the most favored process; the inversion substitution mechanism, in which the H atom approaches the carbon atom from behind one of the C—H bonds, is also available. A CNDO/2 study of dimerization of singlet methylene [119] has led to results in substantial agreement with previous EH results [78], both in the prediction of a high energy barrier for the least-motion approach, and in the description of the least-energy path. For two triplet methylenes it was found, using multi-configuration S.C.F. theory [120], that the least-motion approach is highly favored by the absence of an energy barrier. Other CNDO/2 results are the qualitatively correct prediction on directional and reactivity effects of substituents upon electrophilic aromatic substitution [121], the description of the geometry and electronic structure of the transition state for the Cope rearrangement of endo-tricyclo [5,2,1,0^{2,6}] deca-4,8-diene [122], the evaluation of the influence of solvation on the activation energy for the reaction $CH_3F + F^-$ [123], and the calculation of proton affinities for a number of acids [125]. The CNDO/2 method leads to good correlation with values obtained by acetolysis rates of substituted benzyl and polycyclic arylmethyl p-toluene sulfonates [126], and to an understanding of the stereochemical behavior of bicyclo [2.1.0] pentane interacting with unsaturated molecules [127]. Calculations were made for some models of the activated complex in the thermal rearrangement of bicyclobutane [128], for the two stable conformations of N benzylidene aniline and for the activated complex for their interconversion [129], for the barriers to syn—anti isomerization in imines [130], and for the ring puckering potential in cyclobutane [131].

Before mentioning some results obtained by the MINDO technique, it is worthwhile recalling the fundamental difference in philosophy between this method and the similar CNDO and, in particular, INDO methods. In the latter the parametrization was carried on, trying to reproduce for some simple molecules the results of nonempirical SCF calculations. In the former method the choice of parameters aimed to find the best fit with experimental data [132].

The fascinating field of carbene chemistry has been the object of fruitful MINDO investigations. The structure of the parent carbene has been studied [133] with good results for the geometries and the energy separation of the triplet ground state and the first excited singlet state. Reaction with ethylene was also investigated for both singlet and triplet carbene. The approach of triplet carbene to ethylene was studied by means of the reaction coordinate $r = a_1 - a_2 + 1$ (Fig. 22). The

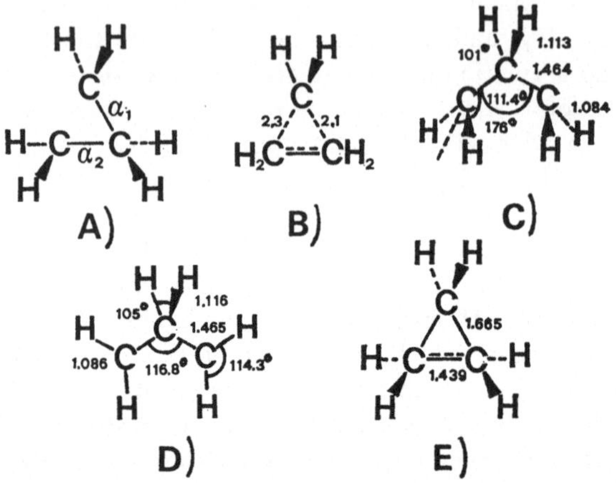

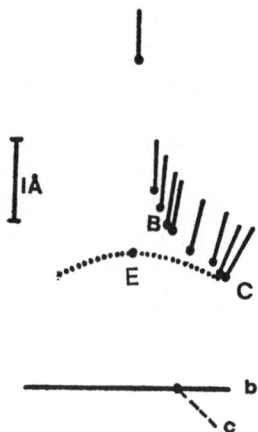

Fig. 22A-E. The $C_2H_4 + CH_2$ (triplet) system: A) coordinates; B) activated complex for the addition; C) biradical at the end of addition; D) diradical ground state; E) activated complex for interconversion (Interatomic distances in Å)

preferred mode of approach turned out to have a plane of symmetry passing through the three carbon atoms and bisecting the methylene groups. The transition-state geometry is also shown in Fig. 22, and the

Fig. 23. Ethylene + triplet carbene: the reaction path; b and c indicate the position of the hydrogen atoms at stages B and C

activation barrier in 5 kcal/mole. Two conformations of the products (C and D) are of practical interest: one is derived from the other by rotation of terminal methylene groups through 90°. Structure D is more stable than C by 3 kcal/mole. The product biradical can undergo scrambling of methylene groups: the π-complex intermediate E, however, is less stable than C by 36 kcal/mole: such an activation energy ought to prevent the interconversion observed in practice. The geometry of the reaction path is shown in Fig. 23. It can be seen that there are three valleys in the potential surface.

Two valleys were found in the potential surface for the reaction of triplet carbene to methane. They correspond to the modes of approach indicated in Fig. 24. The mode leading to 2 CH_3 (b in the figure) is preferred with a barrier height of 3.8 kcal/mole.

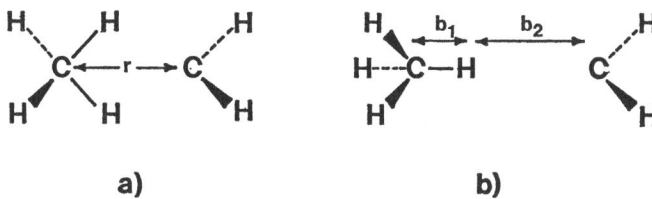

a)　　　　　　　　　　　b)

Fig. 24a and b. The approach of triplet carbene to methane; a) direct insertion; b) abstraction. Reaction coordinate for abstraction $r = b_1 - b_2 + 2$

Experimental results are in greement with these findings. Calculations were also performed for the reaction of singlet methylene, with both ethylene and methane: paths in which no activation energy is involved can be found in both cases. MINDO has been used to study the structure of methyl carbene, cyclohexilidene, formylcarbene and their rearrangements [134] according to the reactions:

$$CH_3-\ddot{C}H \longrightarrow CH_2=CH_2 \qquad \diagdown\!\!\!\!\diagup{}^{\displaystyle\cdot} \longrightarrow \diagdown\!\!\!\!\diagup \qquad OCH-\ddot{C}H \longrightarrow O=C=CH_2$$

A modified form of MINDO was used to discuss, the 1,3 hydrogen shift in propene and the suprafacial and antarafacial 1,5 sigmatropic hydrogen shifts in cis-piperylene: the calculated activation energies were 49.2, 28.3 and 37.0 kcal/mole, respectively [135].

Other applications are studies of: rotation about double bonds [136], ring inversion in cyclooctatetraene [137], the insertion of carbon into ethylene and trans-2-butene to give allenes [138], the barrier height to inversion of nitrogen in hydrazine and alkylamines [139], the Cope rearrangement in 1,5-hexadiene, bullvalene, barbaralane and semibull-valene [140,141], and allowed and forbidden electrocyclic reactions [142].

MINDO has also been used as a basis for a theoretical discussion of ketene cycloadditions [143] of cyclobutene-butadiene isomerization [144].

We now move on to methods belonging to the so-called nonempirical or ab initio category. Most of these methods are based on the self-consistent-field molecular-orbital theory in its LCAO approximation [87]. They can use as basis orbitals Slater-type or Gaussian-type functions. The basis set may further be minimal or extended. Perhaps the simplest and fastest (it is only about 30 times slower than CNDO) is the so-called STO-3G method [145]. Each Slater-type atomic orbital is represented by a linear combination of 3 Gaussian functions and a minimal basis ($1s$, $2s$, $2p$) is used for molecules made of first-row atoms. Exponent values are optimized. The method has been used for the study of the electronic structure of simple organic molecules and ions [146,147] and has proved very successful in calculating molecular geometry; thus it is quite reliable in predicting geometries of species inaccessible to experiment.

As an example, we report the result for the species $C_6H_7^+$ [148]. The ion formed upon protonation of benzene has been given special consideration since it is presumably implicated in electrophilic aromatic substitution reactions [149]. Four possible structures for this ion have been investigated, shown in Fig. 25. The most stable is I, which has the

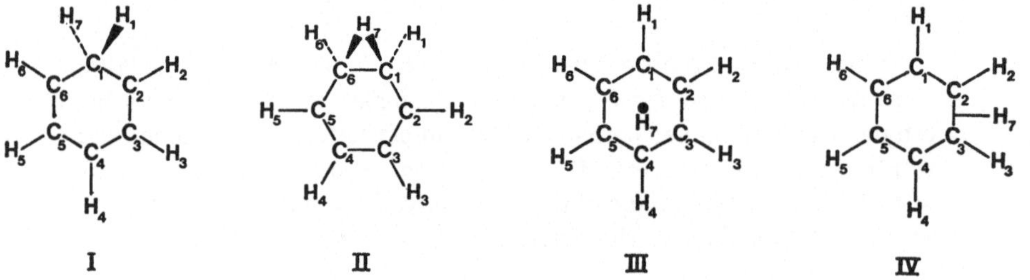

Fig. 25. Structures for the $C_6H_7^+$ ion

following values for the most interesting geometrical parameters (C_{2v} symmetry was assumed):

$$r(C_1-H_1) = 1.106 \text{ Å} \qquad C_1 \hat{C}_2 C_6 = 110.9°$$
$$r(C_2-H_2) = 1.094 \text{ Å} \qquad H_1 \hat{C}_1 H_7 = 105.3°$$
$$r(C_1-C_2) = 1.472 \text{ Å} \qquad H_2 \hat{C}_2 C_3 = 118.3°$$

The next most stable structure is II, with the following geometrical parameters:

$$r(C_1-C_2) = 1.414 \text{ Å} \qquad C_2 \hat{C_1} H_1 = 121.8°$$
$$r(C_1-C_6) = 1.451 \text{ Å} \qquad \beta = 93.9°$$
$$r(C_1-H_7) = 1.333 \text{ Å} \qquad \alpha = 93.4°$$
$$r(C_1-H_1) = 1.090 \text{ Å}$$

where β is the angle between the ring and $C_1H_7C_6$ planes and α is the angle between the planes $C_1H_7C_6$ and $C_1H_1H_6$. The difference in energy between the benzenium and benzenonium ions is 27.7 kcal/mol, or 20.6 kcal/mol if (for the best geometries) calculations are performed with STO-4-31G basis. In this set each inner-shell orbital is represented by a single basis function, taken as the sum of 4 gaussians, and each valence orbital is split into inner and outer parts, described by 3 and 1 gaussian functions, respectively [150]. It should be noted that calculations for the $C_6H_7^+$ ion had been performed earlier, by means of the CNDO/2 technique [151]. Structures I and II had been investigated and the optimized geometries are very close to those found by Pople, but the order of stability is reversed, structure II being more stable by some 20 kcal/mole, as a consequence of the tendency of the CNDO method to favor bridged structures [152].

Other problems dealt with by the Hehre-Pople method are internal rotation in vinylcyclopropane and vinylcyclobutane [153], the structure of homoallyl cation [154] and ethylenebenzenium cation [155], torsional barriers in p-substituted phenols [156], inversion barriers in p-substituted anilines [157], the effects of α-substitution in keto-enol tautomerism [158] and the circumambulatory rearrangement in bicyclo [3.1.0] hex-3-en-2-yl cation [159]:

and in homotropylium ion.

Other minimal basis sets have been used in S.C.F. calculations for chemical reactions. I must mention a very accurate study of the transition state for the geometrical isomerization of cyclopropane, and a search for nondynamical pathways on the potential surface for this reaction when a minimal basis set of Slater orbitals was used [160-162]. Calculations with a basis of this kind were also performed for the coplanar decomposition of cyclobutane [163]. Minimal basis sets based on gaussian functions were used in the study of proton exchange in sulfoxides with retention or inversion of configuration [164] and of the $S + C_2H_4$ reaction path [165]. Extended basis sets, in terms of gaussian functions, were used in S.C.F. studies of the potential energy for the reactions $Cl + H_2 \rightarrow ClH + H$ [166] and $CH_3NC \rightarrow CH_3CN$ [167].

In other cases, as for example in the study of *cis-trans* isomerization of glyoxal [168], the basis was constructed upon gaussian lobe functions, as prescribed by Whitten [169].

In the study of the reaction $CH_3F + F^-$ by the S.C.F. ab initio method using a gaussian type basis set, it was found that, when only s and p functions were used, the transition state was more stable than the reactants [170]. Only the inclusion of polarization factors leads to the expected results [171–173].

The extension of the basis can improve wave functions and energies up to the Hartree-Fock limit, that is, a sufficiently extended basis can circumvent the LCAO approximation and lead to the best molecular orbitals for ground states. However, this is still in the realm of the independent-particle approximation [175], and the use of single Slater-determinant wave functions in the study of potential surfaces implies the assumption that correlation energy remains approximately constant on that part of the surface where reaction pathways develop. In cases when this assumption cannot be accepted, extensive configuration interaction (CI) must be included. A detailed comparison of SCF and CI results is available for the potential energy surface for the reaction $F + H_2 \rightarrow FH + H$ [176].

The basis set included two $1s$, $2s$ and $2p$ functions on fluorine, and two $1s$ functions on each hydrogen atom. Each orbital was obtained as a linear combination of gaussian functions. Linear ($\theta = 0°$) and nonlinear geometries ($\theta = 10, 30, 50, 70, 90°$) were considered (Fig. 26). Single

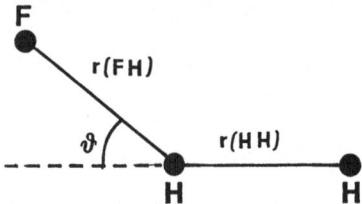

Fig. 26. Coordinates for the reaction $F + H_2$

configuration and 214 configuration wave functions were calculated, for 150 linear geometries and for 40 geometries for each value $\theta \neq 0$. For the linear approach the SCF barrier height and exothermicity are 34.3 and -0.6 kcal/mole. The corresponding CI results are 5.7 and 20.4 kcal/mole, to be compared with the experimental values of 1.7 and 31.2 kcal/mole. The saddle point is at $r(F—H) = 1.06$ Å and $r(H—H) = 0.81$ Å in the SCF calculation; the corresponding CI values are 1.35 and 0.81 Å. Both calculations predict a rather early saddle point: the CI contour map is shown in Fig. 27. Table 4 shows saddle-point geometries and energies as a function of θ.

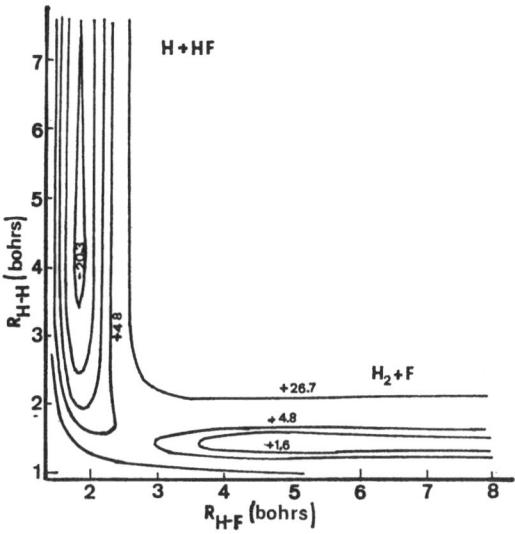

Fig. 27. Contour map for the $F + H_2 \rightarrow FH + H$ reaction (collinear approach)

Table 4. Saddle-point geometries and energies

θ	$r(F—H)$ (a. u.)	$r(H—H)$ (a. u.)	E (kcal/mole) relative to $F + H_2$
0°	2.58	1.54	5.72
10°	2.55	1.54	5.74
30°	2.50	1.56	6.08
50°	2.44	1.60	7.31
70°	2.39	1.68	10.43
90°	2.35	1.87	17.52

It appears that the minimum energy path is linear. According to various authors, the potential surface is not accurate enough to warrant scattering calculations. Calculations including polarization functions (d orbitals on F and p orbitals on H) predict a barrier height of 1.64 kcal/mole and an exothermicity of 34.4 kcal/mole. The minimum energy path obtained in these calculations is qualitatively similar to the one described above.

Ab initio calculations including configuration interactions were reported for the addition of the NH_2 radical to ethylene [177], for dissociation of formaldehyde into radicals [178] and molecular products [179], and for the electrocyclic transformation between cyclobutene and butadiene [180].

An alternative to the MO method for the quantum mechanical treatment of molecular systems is the so-called Valence-Bond (VB) theory where molecular wavefunctions are obtained as linear combinations of covalent and ionic structures. It was shown long ago [181] that for distances larger than equilibrium distances, VB approximate wave functions should be better than MO functions of the same level, and hence VB theory should find its most profitable application in the evaluation of potential surfaces and reaction paths. Although true in principle, this statement has little influence in practice; this is mostly because VB theory has only recently been formulated in a nonempirical form [182–184] so that applications are only just beginning to appear.

The valence-bond method has been used in an ab initio study of the potential surface for H_4^+ [186]. The basis orbitals were linear combinations of gaussian functions with a polarization factor. Calculations were performed with and without inclusion of ionic structures, and with and

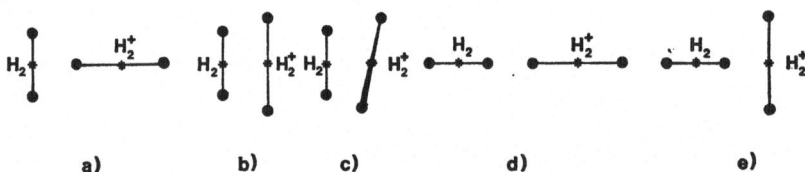

Fig. 28a-e. Five different modes of approach between H_2 and H_2^+. Energy was calculated as a function of R, the distance between the two asterisks, while distances in H_2 and H_2^+ were held constant at their equilibrium value

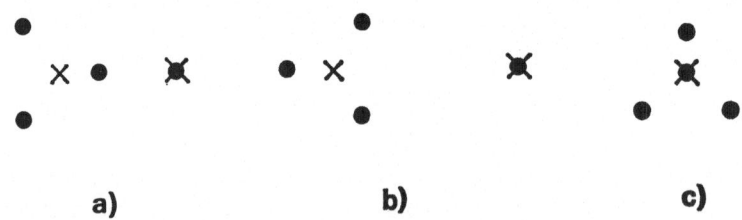

Fig. 29a-c Geometries for the $H_3^+ + H$ system. In c) the central atom is above the plane of the three vertex atoms. H_3^+ is hold in the optimum geometry

without polarization. At each geometry, energy was minimized with respect to orbital exponents, polarization, and linear coefficients. Some asymptotic geometries were considered first, corresponding to $H_2 + H_2^+$ and to $H_3^+ + H$. Various attitudes of approach between H_2 and H_2^+ were then considered, as shown in Fig. 28. The potential between H_3^+ and H was considered next, in geometries shown in Fig. 29. If distances in the fragments H_2 and H_2^+ are allowed to vary, a bound intermediate H_4^+ is found, with dissociation energy 360 cm^{-1}. H_4^+ is an electrostatic complex between H_3^+ and H, with H_3^+ not significantly changed by the H ligand. Potential surfaces were obtained by the VB approach for the reactions $H_2 + D$ [184], LiH + H [187], and Li + F_2 [188]. VB functions were also used in a detailed discussion of symmetry control in the photo-chemical reaction butadiene → cyclobutene [189].

We turn now to reaction dynamics. The theoretical approach to this problem relies on the calculation of trajectories on the potential surface for the reactions. Crossed molecular beams, infrared chemiluminescence and chemical laser techniques are now available to obtain the information experimentally. In particular, it is now possible for simple gas-phase reactions to be studied by both theory and experiment, yielding information on the partitioning of the heat of reaction into vibrational, rotational and translational degrees of freedom. The theoretical approach is at the moment restricted to triatomic systems, or to systems for which a 3-particle model can be constructed. Usually semiempirical potential functions are used, the most popular being derived by the London-Eyring-Polanyi-Sato procedure [190], with an adjustable parameter that allows the barrier height for the reaction to reproduce the value obtained from the experimental activation energy. The Born-Oppenheimer and adjabatic approximations are assumed. Classical mechanics is used to determine the motion of atoms, on the assumption that tunneling is unimportant. The classical equations of motion are solved by numerical integration for a large number of initial values of the dynamical variables. For the reaction A + BC, each trajectory is determined by the value of the initial distance ϱ between A and the center of mass of BC (to be chosen large enough so that there is practically no interaction between the atom and the molecule), the initial relative velocity V, the impact parameter b, the orientation of the molecule, the initial intermolecular distance r_{BC}, and the rotational and vibrational quantum numbers J and v. The end of the trajectory occurs when the distance between any one of the atoms and the center of mass of the other two is large enough. Final values are obtained for the relative velocity of the atom, for the intermolecular distance in the molecule, for the component of its angular momentum, and for the unquantized vibration–rotation energy. Non-reactive (A + BC → A + BC) or reactive (A + BC → AB + C or AC + B) trajectories occur. Dissociative collisions $(A + BC \rightarrow A + B + C)$ require high total energy and are usually outside the explored region. The number of calculated trajectories must be large enough to allow the determination of the reaction probability P_r (given by the ratio of the number of reactive trajectories and trajectories) as a function of V, b, J, v. For fixed values of these variables, the initial value of the remaining variables is chosen at random and the results are averaged by a Monte Carlo procedure. The total cross-section S_r is then obtained from the integral

$$S_r(V, J, v) = 2 \pi \int_0^{b\,\text{max}} P_r(V, J, v, b)\, \text{bdb}$$

where b_{max} is such that P_r is zero for $b > b_{\text{max}}$. The reaction rate can then be calculated by integration or summation over the distribution functions for V, J, v that are characteristic of the physical conditions of the system (*e.g.* the temperature for thermal reactions in bulk gas.)

Calculations of this type have been carried out for the reactions of alkali atoms with methyl iodide [191], in discussing isotope effects on the abstraction and substitution processes for reactions $T + CH_4$ and $T + CD_4$ [192], in the classical $H + H_2$ reaction [193] for which typical

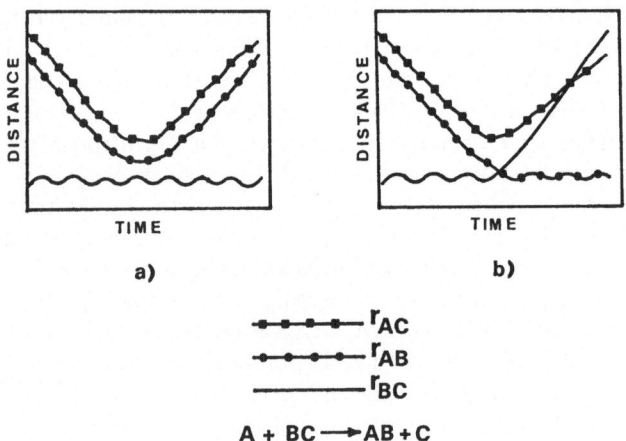

Fig. 30a and b. Typical trajectories for $H + H_2$: a) nonreactive; b) reactive

trajectories are shown in Fig. 30, and for collinear high-energy trajectories in $H + H_2$, where the formation of a complex has been suggested [194]. For the endothermic reaction $H + HF \rightarrow H_2 + F$, it was found that vibrational energy is more effective than translational energy in favoring the reaction [195]; retroverse reaction $F + H_2 \rightarrow H + HF$ has also been studied [190], together with reactions $F + HD$ and $F + D_2$. A small cross section was found, due to back-scattering collisions, and energy partitioning in the products gives high vibrational and low rotational energies.

The position of the downhill part of the energy profile in the entry valley or in the exit valley (Fig. 31) of the potential surface is important in determining release of attractive, mixed, or repulsive energy [197,198].

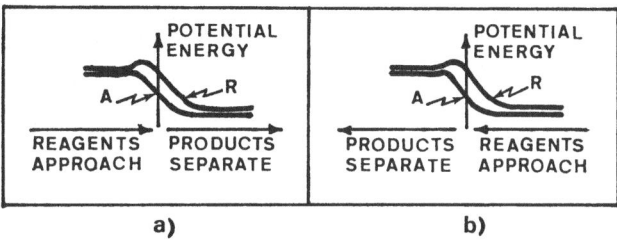

Fig. 31a and b. Potential energy profiles along the reaction coordinate:
a) exothermic b) endothermic processes

The use of the same potential surface has allowed a detailed comparison of results obtained by trajectory calculations and by calculations using the transition-state model [199]. In the $H + H_2$ reaction, the two assumptions incorporated in transition-state theory, *i.e.* equilibrium between transition state and reactants, and transmission coefficient equal to unity, seem to hold very well, but in the collinear $H + HBr \rightarrow H_2 + Br$ reaction the transmission coefficient is found to be less than one, while in the reverse reaction $H_2 + Br \rightarrow H + HBr$ the equilibrium condition is not satisfied.

IV. Summary and Conclusions

Owing to space restrictions, the present review covers by no means all important theoretical work, and there is almost no reference to experimental work, even that directly connected with theoretical results. However, many different theoretical approaches to the chemical reactivity problem have been reviewed. They encompass qualitative deductions from the symmetry of orbitals, configurations and states; empirical and nonempirical calculations of energy and geometry of transition states and of potential energy surfaces at various levels of sophistication, and trajectory calculations. Each method has its own merits and drawbacks and is suitable for a particular range of applications. Whenever possible, the use of more than one method in the study of the same problem turns out to be rewarding. However, even when the most efficient methods are applied to the simplest systems, it is not possible to calculate exact reaction rates. Chemical reactivity is a very complex phenomenon and the present methods of study still have to approach it by means of over-simplified models. Assumptions, such as the validity of the transition-state theory, constancy of solvent or entropy effects, etc. are usually needed. However, if we look at the state of the art ten years ago, we see that progress of our understanding of reactivity has been impressive. It is safe to predict that in the next decade we will do even better, due to improvements in theory, computing facilities, and experimental techniques.

V. Bibliography

[1] A superb presentation of this subject can be found in the book: Woodward, R. B., Hoffmann, R.: The conservation of orbital symmetry. Weinheim: Verlag Chemie 1970.
[2] Woodward, R. B., Hoffmann, R.: J. Am. Chem. Soc. *87*, 395 (1965).
[3] Longuet-Higgins, H. C., Abrahamson, E. W.: J. Am. Chem. Soc. *87*, 2045 (1965).
[4] Hoffmann, R., Woodward, R. B.: J. Am. Chem. Soc. *87*, 2046 (1965).
[5] Ref. 1, page 65.
[6] Hoffmann, R., Woodward, R. B.: J. Am. Chem. Soc. *87*, 4388 (1965).
[7] Woodward, R. B., Hoffmann, R.: J. Am. Chem. Soc. *87*, 2511 (1965).
[8] Holder, R. W.: quoted in Ref. [9].
[9] Berson, J. A.: Accounts Chem. Res. *5*, 406 (1972).
[10] Berson, J. A., Salem, L.: J. Am. Chem. Soc. *94*, 8917 (1972).
[11] For a discussion of the electronic properties of diradicals see: Salem, L., Rowland, C.: Angew. Chem. Intern. Ed. Engl. *11*, 92 (1972).
[12] Baldwin, J. E., Andrist, A. H., Pinschmidt, R. K., Jr.: Acc. Chem. Res. *5*, 402 (1972).
[13] Quoted in: Havinga, E., Schlatmann, J. L. M. A.: Tetrahedron *16*, 151 (1961).
[14] Anastassiou, A. G.: Chem. Commun. *1968*, 15.
[15] Schenneman, E. C. W., Laidlaw, W. G.: J. Am. Chem. Soc. *93*, 5731 (1971).
[16] Kearns, D. R.: J. Am. Chem. Soc. *91*, 6554 (1969).
[17] Eaton, D. R.: J. Am. Chem. Soc. *90*, 4272 (1968).
[18] Chan-Cheng Su: J. Am. Chem. Soc. *93*, 5633 (1971).
[19] Silver, D. M.: Chem. Phys. Letters *14*, 105 (1972).
[20] Fukui, K.: Tetrahedron Letters *1965*, 2009.
[21] Fukui, K.: Bull. Chem. Soc. Japan *39*, 498 (1966).
[22] Fukui, K.: Acc. Chem. Res. *4*, 57 (1971).
[23] See also Pearson, R. G.: Acc. Chem. Res. *4*, 152 (1971).
[24] Pearson, R. G.: J. Am. Chem. Soc. *94*, 8287 (1972).
[25] Lauglet, J., Malrieu, J. P.: J. Am. Chem. Soc. *94*, 7254 (1972).

26) Goddard III, W. A.: J. Am. Chem. Soc. *94*, 793 (1972).
27) Dewar, M. J. S.: Angew. Chem. Intern. Ed. Engl. *10*, 761 (1971).
28) a) Gill, G. B.: Quart. Rev. *22*, 338 (1968);
    b) Sebach, D.: Fortschr. Chem. Forsch. *11*, 177 (1969).
29) Hoffmann, R., Woodward, G. B.: Acc. Chem. Res. *1*, 17 (1968).
30) Caserio, M. C.: J. Chem. Educ. *48*, 782 (1971).
31) Perrin, C. L.: Chem. Brit. *8*, 163 (1972).
32) Salem, L.: Chem. Phys. Letters *3*, 99 (1969).
33) Salem, L., Wright, J. S.: J. Am. Chem. Soc. *91*, 5947 (1969).
34) Pearson, R. G.: J. Am. Chem. Soc. *91*, 4947 (1969).
35) Zimmermann, H. E.: Acc. Chem. Res. *4*, 272 (1971).
36) For a discussion of Möbius systems see: Heilbronner, E.: Tetrahedron Letters *1964*, 1923.
37) Zimmermann, H. E.: Acc. Chem. Res. *5*, 393 (1972).
38) Epiotis, N. D.: J. Am. Chem. Soc. *94*, 1924, 1935, 1941, 1946 (1972).
39) George, T. F., Ross, J.: J. Chem. Phys. *55*, 3851 (1971).
40) Brown, R. D.: In: Molecular orbitals in chemistry, physics and biology (ed. P. O. Löwdin and B. Pullman), p. 495. New York: Academic Press 1964.
41) Fukui, K.: In: Molecular orbitals in chemistry, physics and biology (ed. P. O. Löwdin and B. Pullman), p. 495. New York: Academic Press 1964.
42) Greenwood, H. H., Mc Weeny, R.: Advan. Phys. Org. Chem. *4*, 73 (1966).
43) Zahradnik, R.: Colloques Inter. du CNRS (ed. R. Daudel and A. Pullman). Menton, July 1970.
44) Simonetta, M.: in the press.
45) Dewar, M. J. S.: In: Advances in Chemical Physics VIII, p. 65 (ed. R. Daudel). New York: Interscience Publ. 1965.
46) Hudson, R. F.: Angew. Chem. Intern. Ed. Engl. *12*, 36 (1973).
47) Herndon, W. C.: Chem. Rev. *72*, 157 (1972).
48) Klopman, G.: J. Am. Chem. Soc. *86*, 4550 (1964).
49) Klopman, G.: J. Am. Chem. Soc. *87*, 3300 (1965).
50) Klopman, G., Hudson, R. F.: Theoret. Chim. Acta *8*, 165 (1967).
51) Dougherty, R. C.: J. Am. Chem. Soc. *93*, 7187 (1971).
52) Rastelli, A., Pozzoli, A. S., Del Re, G.: Perkin Trans II *1972*, 1572.
53) Tee, O. S., Yates K.: J. Am. Chem. Soc. *94*, 3079 (1972).
54) Basilevski, M. V., Chlenov, I. E.: Theoret. Chim. Acta *15*, 174 (1969).
55) Basilevski, M. V., Tikhomirov, V. A., Chlenov, I. E.: Theoret. Chim. Acta *23*, 75 (1971).
56) Salem, L.: J. Am. Chem. Soc. *90*, 543, 553 (1968).
57) Devaquet, A., Salem, L.: J. Am. Chem. Soc. *91*, 3793 (1969).
58) Salem, L.: J. Am. Chem. Soc. *95*, 94 (1973); see also Ref. 59).
59) Salem, L., Hoffmann, R., Otto, P.: Proc. Natl. Acad. Sci. U.S. *70*, 531 (1973).
60) Scrocco, E.: This volume, pp. 95-170
61) For a previous review see: Simonetta, M.: Pure Appl. Chem. Suppl. 23rd Congress *1*, 127 (1971).
62) Lifson, S., Warshel, A.: J. Chem. Phys. *49*, 5116 (1968).
63) Warshel, A., Lifson, S.: J. Chem. Phys. *53*, 582 (1970).
64) Bartell, L. S.: J. Chem. Phys. *32*, 827 (1960).
65) Warshel, A., Karplus, M.: J. Am. Chem. Soc. *94*, 5612 (1972).
66) Williams, J. E., Stang, P. J., v. Schleyer, P.: Ann. Rev. Phys. Chem. *19*, 531 (1968).
67) Sheraga, H. A.: Advan. Phys. Org. Chem. *6*, 103 (1968).
68) Gramaccioli, C. M., Simonetta, M.: Acta Cryst. *B 28*, 2231 (1971).

69) Gavezzotti, A., Mugnoli, A., Raimondi, M., Simonetta, M.: Perkin Trans. II *1972*, 925.

70) Bingham, R. C., v. Schleyer, P.: J. Am. Chem. Soc. *93*, 3189 (1971).

71) Gleicher, G. J., v. Schleyer, P.: J. Am. Chem. Soc. *89*, 582 (1967).

72) Simonetta, M., Favini, G., Mariani, C., Gramaccioni, P.: J. Am. Chem. Soc. *90*, 1280 (1968).

73) Beringhelli, T., Gavezzotti, A., Simonetta, M.: J. Mol. Struct. *12*, 333 (1972).

74) Hoffmann, R.: J. Chem. Phys. *39*, 1397 (1963).

75) Dobson, R. C., Hayes, D. M., Hoffmann, R.: J. Am. Chem. Soc. *93*, 6188 (1971).

76) Hoffmann. R.: J. Am. Chem. Doc. *90*, 1475 (1968).

77) Hoffmann, R., Wan, C. C., Neagu, V.: Mol. Phys. *19*, 113 (1970).

78) Hoffmann, R., Gleiter, R., Mallory, F. B.: J. Am. Chem. Soc. *92*, 1460 (1970).

79) Hoffmann, R., Swaminathan, S., Odell. B. G., Gleiter, R.: J. Am. Chem. Soc. *92*, 7091 (1970).

80) Hoffmann, R., Stohrer, W.: J. Am. Chem. Soc. *13*, 6941 (1971).

81) Stohrer, W., Hoffmann, R.: J. Am. Chem. Soc. *94*, 779 (1972).

82) Stohrer, W., Hoffmann, R.: J. Am. Chem. Soc. *94*, 1661 (1972).

83) see *e.g.*: Feler, G.: Theoret. Chim. Acta *12*, 412 (1968).

84) Csizmadia, I. H., Gunning, H. E., Kosavi, R. K., Strausz, O. P.: J. Am. Chem. Soc. *95*, 133 (1973).

85) Farkas, M., Kruglyac, J. A.: Nature *223*, 523 (1969).

86) Phelan, N. F., Jaffé, H. H., Orchin, M.: J. Chem. Educ. *44*, 626 (1967).

87) Roothaan, C. C. J.: Rev. Mod. Phys. *23*, 69 (1951).

88) Pariser, R. Parr, R. G.: J. Chem. Phys. *21*, 466, 767 (1953).

89) Pople, J. A.: Trans. Faraday Soc. *49*, 1375 (1953).

90) Pople, J. A., Santry, D. P., Segal, G. A.: J. Chem. Phys. *43*, 5129 (1965).

91) Dewar, M. J. S., Klopman, G.: J. Am. Chem. Soc. *89*, 3089 (1967).

92) Pople, J. A., Beveridge, D. L., Dobosh, P. A.: J. Chem. Phys. *47*, 2026 (1967), see also Ref. 93).

93) Dixon, R. N.: Mol. Phys. *12*, 83 (1967).

94) Baird, N. C., Dewar, M. J. S.: J. Chem. Phys. *50*, 1262 (1969).

95) Dewar, M. J. S., Haselbach, E.: J. Am. Chem. Soc. *92*, 590 (1970).

96) Pople, J. A., Beveridge, D. L.: Approximate molecular orbital theory. New York: McGraw-Hill Book 1970.

97) Klopman, G., O'Leary, B.: All-valence electrons SCF calculations. Berlin–Heidelberg–New York: Springer 1970.

98) Murrel, J. N., Harget, A. J.: Semi-empirical self-consistent-field molecular-orbital theory of molecules. London: Wiley-Interscience 1972.

99) Pople, J. A., Segal, G. A.: J. Chem. Phys. *44*, 3289 (1966).

100) Gamba, A., Morosi, G., Simonetta, M.: Chem. Phys. Letters *3*, 20 (1969).

101) Rutledge, R. M., Saturno, A. F.: J. Chem. Phys. *43*, 597 (1965).

102) Allen, L. C.: In: Quantum theory of atoms, molecules and solid state (ed. P. O. Löwdin), p. 62. New York: Academic Press 1966.

103) Yonezawa, T., Nakatsuji, H., Kato, H.: J. Am. Chem. Soc. *90*, 1239 (1968).

104) Olah, G. A., Klopman, G., Schlosberg, R. H.: J. Am. Chem. Soc. *91*, 3261 (1968).

105) Gole, J. L.: Chem. Phys. Letters *3*, 577 (1969); *4*, 408 (1969).

106) Kollmar, H., Smith, H. O.: Chem. Phys. Letters *5*, 7 (1970); see also Ref. 107).

107) Allinger, N. L., Tai, J. C., Wu, F. T.: J. Am. Chem. Soc. *92*, 579 (1970).

108) Lathan, W. A., Hehre, W. J., Pople, J. A.: J. Am. Chem. Soc. *93*, 6377 (1971).

109) Dyczmons, V., Staemmler, V., Kutzelnigg, W.: Chem. Phys. Letters *5*, 361 (1970); see also Ref. 112).

110) Van Der Lugt, W. Th. A. M., Ros, P.: Chem. Phys. Letters *4*, 389 (1969).
111) Mulder, J. J. C., Wright, J. S.: Chem. Phys. Letters *5*, 445 (1969); see also Ref. 112).
112) Ritchie, C. D., King, H. F.: J. Am. Chem. Soc. *90*, 825 (1968).
113) Grimm, F. A., Godoy, J.: Chem. Phys. Letters *6*, 336 (1970).
114) Ehrenson, S.: Chem. Phys. Letters *3*, 585 (1969).
115) Watson, R. E., Jr., Ehrenson, S.: Chem. Phys. Letters *9*, 351 (1971).
116) Guest, M. F., Murrel, J. N., Pedley, J. B.: Mol. Phys. *20*, 81 (1971).
117) Morokuma, K., Davis, R. E.: J. Am. Chem. Soc. *94*, 1060 (1972).
118) Ehrenson, S., Newton, M. D.: Chem. Phys. Letters *13*, 24 (1972).
119) Tantardini G. F., Simonetta, M.: Israel J. Chem. *10*, 582 (1972).
120) Bash, H.: J. Chem. Phys. *55*, 1700 (1971).
121) Howe, G. R.: Chem. Commun. *1970*, 868.
122) Beltrame, P., Gamba, A., Simonetta, M.: Chem. Commun. *1970*, 1660.
123) Cremaschi, P., Gamba, A., Simonetta, M.: Theoret. Chim. Acta .*25*, 237 (1972); see also Ref. 124).
124) Lowe, J. P.: J. Am. Chem. Soc. *93*, 302 (1971).
125) Rode, B. M., Engelbrecht, A.: Chem. Phys. Letters *16*, 560 (1972).
126) Streitwieser, A., Jr., Hammond, H. A., Jagow, R. H., Williams, R. M., Jesaitis, R. G., Chang, C. J., Wolf, R.: J. Am. Chem. Soc. *92*, 5141 (1970).
127) Collins, F. S., George, J. K., Trindle, C.: J. Am. Chem. Soc. *94*, 3732 (1972).
128) Wiberg, K. B.: Tetrahedron *24*, 1083(1968).
129) Warren, C. H., Wettermark, G., Weiss, K.: J. Am. Chem. Soc. *93*, 4658 (1971).
130) Raban, M.: Chem. Commun. *1970*, 1415.
131) Wright, J. S., Salem, L.: Chem. Commun. *1969*, 1370.
132) Dewar, M. J. S.: Topics in current chemistry, *23*, 1. Berlin—Heidelberg–New York: Springer 1971.
133) Bodor, N., Dewar, M. J. S., Wasson, J. S.: J. Am. Chem. Soc. *94*, 9095 (1972).
134) Bodor, N., Dewar, M. J. S.: J. Am. Chem. Soc. *94*, 9103 (1972).
135) Bingham, R. C., Dewar, M. J. S.: J. Am. Chem. Soc. *94*, 9107 (1972).
136) Dewar, M. J. S., Haselbach, E.: J. Am. Chem. Soc. *92*, 590 (1970).
137) Dewar, M. J. S., Harget, A. J., Haselbach, E.: J. Am. Chem. Soc. *91*, 7521 (1969).
138) Dewar, M. J. S., Hasselbach, E., Shanshal, M.: J. Am. Chem. Soc. *92*, 3505 (1970).
139) Dewar, M. J. S., Shanshal, M.: J. Am. Chem. Soc. *91*, 3654 (1969).
140) Brown, A., Dewar, M. J. S., Schaeller, W.: J. Am. Chem. Soc. *92*, 5516 (1970).
141) Dewar, M. J. S., Lo, D. H.: J. Am. Chem. Soc. *93*, 7101 (1971).
142) Dewar, M. J. S., Kirschner, S.: J. Am. Chem. Soc. *93*, 4290, 4291, 4292 (1971).
143) Sustmann, R., Ansmann, A., Vahrenholt, F.: J. Am. Chem. Soc. *94*, 8099 (1972).
144) Mc Iver, J. W., Jr., Komornicki, A.: J. Am. Chem. Soc. *94*, 2625 (1972).
145) Here, W. J., Stewart, R. F., Pople, J. A.: J. Chem. Phys. *51*, 2657 (1969).
146) Lathan, W. A., Here, W. J., Pople, J. A.: J. Am. Chem. Soc. *93*, 808 (1971).
147) Radom, L., Pople, J. A., Buss, V., Schleyer, P. v. R.: J. Am. Chem. Soc. *93*, 1813(1971).
148) Here, W. J., Pople, J. A.: J. Am. Chem. Soc. *94*, 6901 (1972).
149) Olah, G.: Acc. Chem. Res. *4*, 240 (1971).
150) Ditchfield, R., Here. W. J., Pople, J. A.: J. Chem. Phys. *54*, 724 (1971).
151) Jakubetz, W., Schuster, P.: Angew. Chem. Intern. Ed. Engl. *10*, 497 (1971).
152) Sustman, R., Williams, J. E., Dewar, M. J. S., Allen, L. C., Schleyer, P. v. R.: J. Am. Chem. Soc. *91*, 5350 (1969).
153) Hehre, W. J.: J. Am. Chem. Soc. *94*, 6592 (1972).

154) Hehre, W. J., Hilberty, P. C.: J. Am. Chem. Soc. *94*, 5917 (1972).
155) Hehre, W. J.: J. Am. Chem. Soc. *94*, 5919 (1972).
156) Radom, L., Hehre, W. J., Pople, J. A., Carlson, G. L., Fateley, W. H.: Chem. Commun. *1972*, 308.
157) Hehre, W., Radom, L., Pople, J. A.: Chem. Commun. *1972*, 669.
158) Hehre, W. J., Lathan, W. A.: Chem. Commun. *1972*, 771.
159) Here, W. J.: J. Am. Chem. Soc. *94*, 8908 (1972).
160) Jean, V., Salem, L., Wright, J. S., Horsley, J. A., Moser, C., Stevens, R. M.: Pure Appl. Chem. Suppl. (23rd Congr.) *1*, 197 (1971).
161) Horsley, J. A., Jean, V., Moser, C., Salem, L., Stevens, R. M., Wright, J. S.: J. Am. Chem. Soc. *94*, 279 (1972).
162) Salem, L.: In: The transition state (ed. J. E. Dubois), p. 97. London: Gordon and Breach 1972.
163) Wright, J. S., Salem, L.: J. Am. Chem. Soc. *94*, 322 (1972).
164) Rank, A., Wolfe, S., Csizmadia, I. G.: Can. J. Chem. *47*, 113 (1969).
165) Strausz, O. P. Gunning, H. E., Denes, A. S., Csizmadia, I. G.: J. Am. Chem. Soc. *94*, 8317 (1972).
166) Rothenberg, S., Schaefer III, H. F.: Chem. Phys. Letters *10*, 565 (1971).
167) Liskow, D. H., Bender, C. F., Schaefer III, H. F.: J. Am. Chem. Soc. *94*, 5178 (1972).
168) Ha, T. H.: J. Mol. Struct. *12*, 171 (1972).
169) Whitten, J. L.: J. Chem. Phys. *44*, 359 (1966).
170) Berthier, G., David, D. J., Veillard, A.: Theoret. Chim. Acta *19*, 329 (1969).
171) Dedieu, A., Veillard, A.: Chem. Phys. Letters *5*, 328 (1970).
172) Dedieu, A., Veillard, A.: In: The transition state (ed. J. E. Dubios), p. 153. London: Gordon and Breach 1972.
173) Dedieu, A., Veillard, A.: F. Am. Chem. Soc. *94*, 6730 (1972); see also Ref. 179).
174) Duke, A. J., Bader, R. F. W.: Chem. Phys. Letters *10*, 631 (1971).
175) Löwdin, P. O.: Advan. Chem. Phys. *2*, 207 (1959).
176) Bender, C. F., Pearson, P. K., O'Neil, S. V., Schaefer III, H. F.: J. Chem. Phys. *56*, 4626 (1972).
177) Smith, S., Buenker, R. J., Peyerimhoff, S. D., Michejda, C. J.: J. Am. Chem. Soc. *94*, 7620 (1972).
178) Fink, W. H.: J. Am. Chem. Soc. *94*, 1073 (1972).
179) Fink, W. H.: J. Am. Chem. Soc. *94*, 1078 (1972).
180) Hsu, K., Buenker, R. J., Peyerimhoff, S. D.: J. Am. Chem. Soc. *94*, 5639 (1972).
181) Slater, J. C.: J. Chem. Phys. *19*, 220 (1951).
182) Simonetta, M., Gianinetti, E., Vandoni, I.: J. Chem. Phys. *48*, 1579 (1968); see also Ref. 185).
183) Harrison, J. F., Allen, L. C.: J. Am. Chem. Soc. *91*, 807 (1969).
184) Ladner, R. C., Goddard III, W. A.: J. Chem. Phys. *51*, 1073 (1969).
185) Raimondi, M., Simonetta, M., Tantardini, G. F.: J. Chem. Phys. *56*, 5092 (1972).
186) Poshusta, R. D., Zetik, D. F.: J. Chem. Phys. *58*, 118 (1973).
187) Goddard III, W. A., Ladner, R. C.: J. Am. Chem. Soc. *93*, 6750 (1971).
188) Baliut-Kurti, G. G., Karplus, M.: Chem. Phys. Letters *11*, 203 (1971).
189) van der Lugt, W. Th. A. M., Oosterhoff, L. J.: J. Am. Chem. Soc. *91*, 6042 (1969).
190) Sato, S.: J. Chem. Phys. *23*, 592 (1965).
191) Blais, N. C., Bunker, D. L.: J. Chem. Phys. *37*, 2713 (1962).
192) Bunker, D. L., Pattengill, M. D.: J. Chem. Phys. *53*, 3041 (1970).
193) Karplus, M., Porter, R. N., Sharma, R. D.: J. Chem. Phys. *43*, 3259 (1965).
194) Careless, P. N., Hüatt, D.: Chem. Phys. Letters *14*, 358 (1972).

[195] Anderson, J. B.: J. Chem. Phys. *52*, 3849 (1970).
[196] Muckerman, J. T.: J. Chem. Phys. *54*, 1155 (1971).
[197] Kuntz, P. J., Nemeth, E. M., Polanyi, J. C., Rosmer, S. D., Young, C. E.: J. Chem. Phys. *44*, 1168 (1966).
[198] Polanyi, J. C.: Acc. Chem. Res. *5*, 161 (1972).
[199] Karplus, M.: In: The transition state, (ed. J. E. Dubois), p. 279. London: Gordon and Breach 1972.

Received April 2, 1973

Graph Theory and Molecular Orbitals

Dr. Ivan Gutman and Prof. Dr. Nenad Trinajstić

Institute "Rugjer Bošković", Zagreb, Croatia, Yugoslavia

Contents

I.	Introduction and Glossary	50
II.	Elements of Graph Theory	51
	A. Definition of the Graph	52
	B. Graphs and Topology	54
	C. Graphs and Matrices	55
	D. Graphs and Groups	58
	E. Subgraphs	59
	F. Graph Colouring	59
III.	Equivalence between Elementary Molecular Orbital Theory and the Graph Spectral Problem	61
	A. Hückel Theory	61
	B. The Sachs Theorem	63
IV.	The Pairing Theorem	66
V.	Bond Orders and Charges	68
VI.	Unstable π-Electron Systems	72
VII.	The Hückel $(4m+2)$ Rule and its Generalizations	77
	A. Hückel Rule	78
	B. Loop Rule	82
VIII.	Molecular Orbitals and Kekulé Structures	84
IX.	Unsolved Problems and Future Developments	87
X.	References	88

I. Introduction and Glossary

Graph Theory is a very exciting part of pure and applied mathematics because it has a universal character. Thus, Graph Theory is nowdays used in very diverse fields, for example, economics[1], theoretical physics[2,3], psychology[4], biomathematics[5], linguistics[6], nuclear physics[2,3], sociology[7], transportation[8], etc. Similarly, in recent years we have witnessed a remarkable growth in the applications of the principles of Graph Theory in chemistry[9]. There are several reasons for this: Graph Theory has provided a valuable tool with which experimental chemists, using simple rules, have obtained many useful qualitative predictions about the structure and reactivity of various compounds of interest. All these predictions can be made using just pencil and paper without the need for any lengthy and sometimes difficult theoretical calculations; this is, of course, a very attractive feature of Graph Theory. Similarly, Graph Theory can be used as a foundation for the representation and categorization of a very large number of chemical systems[10]. Graph Theory is especially attractive for organic chemists, because it enables the various combinatorial problems of organic chemistry to be solved, *e.g.* the number of structural isomers[11-15], the configurations of annulenes[16], the number of Kekulé structures[17-19], etc.

Graph Theory can also be applied directly to quantum chemistry; a good illustration of this is the graph theoretical derivation of the Pairing Theorem, derived earlier by Coulson and Rushbrooke[20] in a different way.

We discuss here some aspects of the application of Graph Theory to quantum chemistry and examine the close relationship between Graph Theory and elementary molecular orbital (MO) theory. The simple MOs can be obtained by solving the *topological matrix*[21, 22] of the molecule.

The topological matrix is a matrix whose rows and columns are numbered according to the atoms of the molecule and whose elements are non-zero only for bonded atom pairs.

If the molecule is represented by its graph, the adjacency matrix can be associated with the molecular graph.

The adjacency matrix is a matrix whose rows and columns are numbered according to the vertices of the graph and whose elements are non-vanishing for edges only.

It is apparent from the definitions that the topological matrix and adjacency matrix are identical, the former being associated with the molecule and the latter with the molecular graph. Two points must be emphasized here. Firstly, the power of survival of simple molecular orbitals is due to the fact that the topology of the molecule shapes the MOs. Therefore, the simple MOs should correctly be called *"topological molecular orbitals"*, the name originally proposed by Ruedenberg[23], the properties obtained from them being determined solely by molecular topology and independently of the quantum-mechanical model. Topological MOs are exemplified by the Hückel molecular orbitals[24]. Secondly, since the topological and adjacency matrices are identical, Graph Theory may equally well be used for studies of the molecular properties (energy, bond orders, charge distributions) which depend on topology. Furthermore, Graph Theory may be used in such a manner that a number of results can be obtained directly by considering only the molecular topology without going through the mathematical manipulations of solving the eigenvalue problem of the adjacency matrix.

Finally, we give a short glossary to help research chemists understand the language of Graph Theory.

Graph Theory terminology[1]	Chemical terminology
Graph	Structural formula
Vertex	Atom
Edge	Bond
Degree	Valency
Cycle	Ring
Tree	Acyclic hydrocarbon
Chain	Linear hydrocarbon
Bipartite graph	Alternant hydrocarbon
Non-bipartite graph	Non-alternant hydrocarbon
Adjacency matrix A	Topological matrix
Eigenvector of A	Topological MO
Eigenvalue of A	MO energy level
Number of zeros in the spectrum of A	Number of NBMOs (NBMO = non-bonding molecular orbitals)
Characteristic polynomial	Secular equation
Kekulé graph	Kekulé structure

[1] There is a difficulty with graph theoretical terminology. A number of active researchers in this field use their own terms.
It is our intention in the present article to use the terminology of Graph Theory which we propose for standard use in the chemical literature.

II. Elements of Graph Theory

In this section we give a brief survey of the mathematical apparatus of Graph Theory. Since this article is *"application-oriented"*, and thus is designed for the chemical community at large, mathematical rigour is omitted whenever possible.

All necessary definitions[25] will be given in this chapter. The details of Graph Theory can be found in some excellent books[8,26].

A. Definition of the Graph

We will consider a set $\mathscr{V}$ of some elements ("vertices") and a binary relation $\mathscr{A}$ defined on the set $\mathscr{V}$. The meaning of this is that two elements v_1 and v_2 of the set $\mathscr{V}$ (two vertices) either belong to the relation

$$(v_1,\ v_2) \in \mathscr{A} \tag{1}$$

or they do not

$$(v_1,\ v_2) \notin \mathscr{A}. \tag{2}$$

For an arbitrary ordered pair of vertices either Eq. (1) or Eq. (2) is valid. When Graph Theory is used in chemical problems the relation $\mathscr{A}$ has to be symmetric and antireflexive, *i.e.*

$$(v_1,\ v_2) \in \mathscr{A} \rightarrow (v_2,\ v_1) \in \mathscr{A} \tag{3}$$

and

$$(v_1,\ v_2) \in \mathscr{A} \rightarrow v_1 \neq v_2. \tag{4}$$

A graph is defined as an ordered pair G

$$G = (\mathscr{V},\ \mathscr{A}). \tag{5}$$

This abstract definition can be visualized when the vertices are drawn as small circles and two vertices are connected by a line if they belong to relation $\mathscr{A}$ (see Eq. (1)). These connecting lines are called "edges".

Examples

1) $\mathscr{V} = [1,\ 2]$

   $\mathscr{A} = [(1,\ 2),\ (2,\ 1)]$

   $(\mathscr{V},\ \mathscr{A})$ is the graph

2) $\mathscr{V} = [1,\ 2,\ 3,\ 4]$

   $\mathscr{A} = [(1,\ 2),\ (2,\ 1),\ (2,\ 3),\ (2,\ 4),\ (3,\ 2),\ (3,\ 4),$
   $(4,\ 2),\ (4,\ 3)]$

   $(\mathscr{V},\ \mathscr{A})$ is the graph

One property of molecules seems to be very close to a binary relation; namely two atoms in a molecule are either bonded or not bonded. There- fore, molecules can be represented by graphs when the only property considered is the existence or not of a chemical bond. We call this property the *molecular topology*. All other molecular properties (*e.g.* geometry, type of bonding, symmetry, chirality, etc.) are neglected. The analogy between structural formulae and graphs is obvious. Some examples are given in Fig. 1. When conjugated hydrocarbons are con-

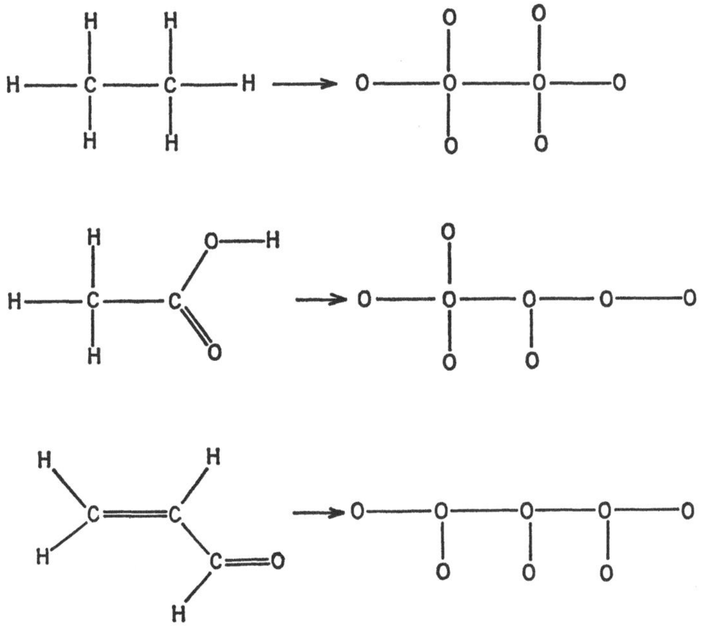

Fig. 1. The graph representation of some formulae

sidered, the related graphs correspond to the carbon—carbon σ—bond skeleton, while the H atoms, π bonds, and C—H σ bonds are neglected.

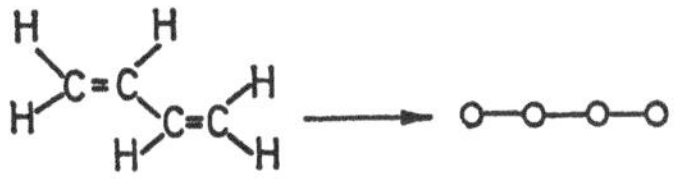

In the present article all graphs are assumed to correspond to conjugated hydrocarbons in this way.

The set $\mathscr{V}$ is supposed to be finite and to have N elements. Thus, N also denotes the number of carbon atoms. The number of edges (C—C bonds) is denoted by ν.

B. Graphs and Topology

A *path* in a graph is an ordered set of edges $(e_1, e_2, \ldots, e_n)$ with the property: the edge e_j $(1 < j \leq n)$ starts from the edge where the edge e_{j-1} ends.

The *length* of such a path is n. Thus, for example, (1, 2, 3), (1, 2, 2, 5), (1), (2, 3, 4, 5), (1, 1, 1, 1, 1) are paths in the graph G, but (1, 6) is not.

If the path ends at the same vertex from which it started, we call it a *loop*. Paths (2, 3, 4, 5) and (2, 3, 3, 2) are loops in the graph G.

The shortest path between two vertices r and s is called the *distance* between two vertices and is denoted as $d(r, s)$. It is not difficult to see that

$$d(r, s) = 0 \text{ if, and only if, } r = s,$$
$$d(r, s) = d(s, r) \text{ and} \tag{6}$$
$$d(r, s) + d(s, t) \geq d(r, t).$$

Obviously, the distance function d has only integral values. According to the definition of the graph

$$d(r, s) = 1 \text{ if, and only if, } (r, s) \in \mathscr{A}. \tag{7}$$

If there is no path between two vertices (*i.e.* $d = \infty$), they belong to different *components* of a graph. All vertices of a graph component have finite distances. Although all graphs corresponding to molecules necessarily have only one component, the number of graph components is an important property and we shall use it later (see Section III.B). The graphs G_1, G_2, and G_3 have one, two, and three components, respectively.

Once the distance function is defined, it is easy to introduce the notion of *neighbours*. All vertices among which the distance is unity are first neighbours (or more simply: neighbours). Second, third, etc. neighbours are defined in a completely analogous way. The number of first, second, third, etc. neighbours of a vertex is denoted as σ_1, σ_2, σ_3, etc. σ_1 is also called the *degree* of the vertex. It can easily be seen that the number of edges starting from a vertex is equal to the degree of the vertex. Hence,

$$\sum \sigma_1 = 2\,\nu \qquad (8)$$

where the summation goes over all the vertices of the graph. Only vertices of degree 1, 2, and 3 appear in the graphs belonging to conjugated hydrocarbons.

Two special types of graphs will be important for further discussion (see Section III):

a) If all vertices are of the degree 2, the graph is called a *cycle*. Graphs G_1 and G_2 in Fig. 2 are cycles.

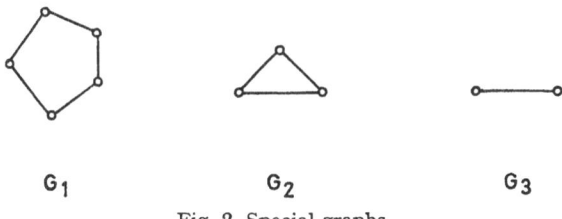

G_1 $\qquad\qquad$ G_2 $\qquad\qquad$ G_3

Fig. 2. Special graphs

b) If the distance between all vertices is unity, the graph is *complete*. The degree of all vertices in such a graph is $\sigma_1 = N-1$, and it is called a *complete graph of degree σ_1*. Graphs G_2 and G_3 in Fig. 2 are complete graphs of degree 2 and 1, respectively. Complete graphs of degree one play a significant role in Sachs' Theorem [27] (see Section III.B).

C. Graphs and Matrices

The following notation is adopted[a]. M^+ denotes the transposed matrix M, i.e.

$$(M^+)_{pq} = M_{2P} \qquad (9)$$

I and O denotes unit and zero matrices, respectively, of an arbitrary type.

There are several ways of assigning a matrix to a graph [8,28], but only the *adjacency matrix* [8] A is important for us. The adjacency matrix is defined as follows

$$A_{rs} = \begin{cases} 1 \text{ if } (r, s) \in \mathscr{A} \\ 0 \text{ if } (r, s) \notin \mathscr{A}. \end{cases} \qquad (10)$$

---

[a] Matrices are presented in heavy print.

The dimension of A is, of course, N. The adjacency matrix is symmetric and has zero diagonal elements

$$A^+ = A \qquad (11)$$

$$A_{rr} = 0 \ (r = 1, 2, \ldots, N) \qquad (12)$$

as a consequence of Eqs. (3) and (4). Because of Eq. (7) the adjacency matrix describes the connectivity of the vertices in the graph.

A_{rs} is also the number of paths of length 1 between vertices r and s. One can see now that

$$A_{rj}A_{js} = \begin{cases} 1 & \text{if there is a path of length 2 between} \\ & r \text{ and } s \text{ and which passes through } j \\ 0 & \text{if there is no such path} \end{cases} \qquad (13)$$

and

$$(A^2)_{rs} = \sum_{j=1}^{N} A_{rj}A_{js} \qquad (14)$$

where this expression represents the number of paths of length 2 between r and s. This consideration can be extended for an arbitrary exponent n, i.e.:

$$(A^n)_{rs} = \frac{\text{number of paths of length } n \text{ between}}{\text{vertices } r \text{ and } s.} \qquad (15)$$

This equation will be utilized in Section VII.

The actual form of the adjacency matrix depends on the numbering of the vertices. Therefore, although the graphs G_1 and G_2 are obviously identical, the corresponding adjacency matrices A_1 and A_2 are not:

$$A_1 = \begin{bmatrix} 0 & 1 & 0 & 0 & 0 & 0 \\ 1 & 0 & 1 & 0 & 0 & 1 \\ 0 & 1 & 0 & 1 & 0 & 0 \\ 0 & 0 & 1 & 0 & 1 & 1 \\ 0 & 0 & 0 & 1 & 0 & 0 \\ 0 & 1 & 0 & 1 & 0 & 0 \end{bmatrix} \quad A_2 = \begin{bmatrix} 0 & 0 & 0 & 0 & 1 & 0 \\ 0 & 0 & 0 & 0 & 1 & 1 \\ 0 & 0 & 0 & 0 & 1 & 1 \\ 0 & 0 & 0 & 0 & 0 & 1 \\ 1 & 1 & 1 & 0 & 0 & 0 \\ 0 & 1 & 1 & 1 & 0 & 0 \end{bmatrix} \tag{16}$$

We have relied on the reader's intuition in recognizing that the graphs G_1 and G_2 above are identical. This procedure for recognizing identical graphs is simple enough for small graphs like G_1 and G_2, but for a general case it remains one of unsolved basic problems of Graph Theory.

Since A is not independent of vertex numbering, other functions which are invariant to the vertex numbering are rather important. For reasons which will become clear in the next section, the graph spectrum is the most important graph invariant. Let C be the *eigenvector matrix* of A, *i.e.*

$$CA = XC \tag{17}$$

where X is a diagonal matrix, whose diagonal elements are called *eigenvalues* of the matrix A. If C and X are of the form

$$C = \begin{bmatrix} C_1 \\ C_2 \\ \vdots \\ C_N \end{bmatrix} \text{ and } X = \begin{bmatrix} x_1 & & & \\ & x_2 & & O \\ & & \ddots & \\ O & & & x_N \end{bmatrix} \tag{18}$$

then

$$C_i A = x_i C_i \quad i = 1, 2, \ldots, N \tag{19}$$

where C_i are *eigenvectors* with the eigenvalue x_i. The set of all eigenvalues $[x_1, x_2, \ldots, x_N]$ is called the *spectrum* of the graph (*i.e. graph spectrum*). There is an important property of the graph spectrum:

$$-\sum \leqslant x_i \leqslant +\sum \tag{20}$$

where $\sum$ is the maximal vertex degree in the graph. In the chemically relevant graphs the whole spectrum lies in the interval from -3 to $+3$ (Ref. [29]). If matrices A_1 and A_2 correspond to two different numberings of the same graph, then there is a permutation Π such that

$$\Pi A_1 \Pi^+ = A_2 \tag{21}$$

Since Π is a unitary matrix

$$\Pi\Pi^+ = I \qquad (22)$$

the matrices A_1 and A_2 are similar and they have, according to a theorem from matrix algebra [30], the same eigenvectors and eigenvalues. Hence, the spectrum is a graph invariant.
Eq. (19) can be rewritten in the form:

$$C_i\,(x_i I - A) = O \qquad (23)$$

and thus the graph spectrum is the set of roots of the *secular equation*:

$$\det |x I - A| = 0. \qquad (24)$$

The polynomial

$$P_G(x) = \det |x I - A| \qquad (25)$$

is called the *characteristic polynomial of the graph* and it is also a graph invariant. $P_G(x)$ is of degree N, *i.e.*:

$$P_G(x) = \sum_{n=0}^{N} a_n\, x^{N-n} \qquad (26)$$

a_n are the coefficients of the characteristic polynomial. The systematic study of graph spectra is a novel problem. The most comprehensive publication available on this topic is Ref. [31].

D. Graphs and Groups

Symmetry properties of molecules cannot be used in a proper way in molecular graphs. This is obvious since G_1 and G_2 equally represent the benzene molecule:

However, Group Theory can be employed in the following manner. There are $N!$ permutations of the graph vertices Π, some of them leaving the adjacency matrix invariant, *i.e.*:

$$\Pi A \Pi^+ = A. \qquad (27)$$

Transformation (27) is called an *automorphism* of the graph. Thus, for example, Π_1 is an automorphism of the graph G:

The set of all automorphisms of a graph forms a group [32-34]. If the graph is drawn so that it possesses some symmetry, every symmetry operation is in fact an automorphism. Therefore, graphs can be treated as if they possess the symmetry of the corresponding molecule.

Using symmetry considerations the eigenvalue problem can be solved for several classess of graphs in closed analytical form [35-49].

In addition, the group of automorphisms can possess some additional elements which do not correspond to any symmetry operation. Therefore, the eigenvalue spectra of conjugated molecules sometimes exhibit higher symmetry than the geometrical symmetry group would admit. Wild, Keller and Günthard [33] have discussed the nature of this excessive symmetry and have given a number of conditions for its occurrence.

E. Subgraphs

If $\mathscr{V}_1$ is a subset of $\mathscr{V}$ and $\mathscr{A}_1$ a subset of $\mathscr{A}$, *i.e.*:

$$\mathscr{V}_1 \subseteq \mathscr{V} \tag{28a}$$

$$\mathscr{A}_1 \subseteq \mathscr{A} \tag{28b}$$

and if $\mathscr{A}_1$ is a relation on the set $\mathscr{V}_1$, the graph $G_1 = (\mathscr{V}_1, \mathscr{A}_1)$ is called a *subgraph of the graph* G. Thus, G_1, G_2, and G_3 are subgraphs of the graph G:

F. Graph Colouring

The problem of colouring a graph with a given number of different colours in such a way that adjacent vertices are always differently coloured is a typical graph theoretical problem. Although the four-colour problem [50] has not yet been solved, the *two-colour problem*, arising in graphs of interest in chemistry, has been solved. Graphs which can be coloured in two colours are called bipartite graphs (or sometimes bichromatic graphs). Fig. 3 shows some bipartite and nonbipartite graphs, the colouring process being indicated by stars (*) and circles (O). Vertices

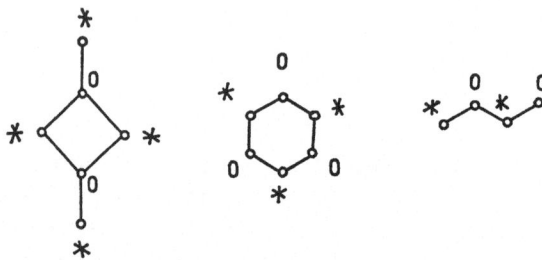

Fig. 3. (a) Bipartite graphs (no odd-membered rings)

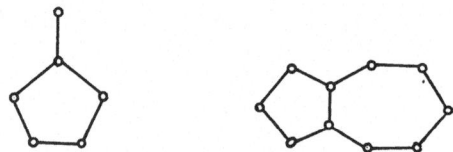

Fig. 3. (b) Non-bipartite graphs (odd-membered rings)

of different "colour" in bipartite graphs will therefore be called *starred* and *unstarred*. Conventionally

$$p \geqslant q \tag{29}$$

where p and q denote the number of starred and unstarred vertices respectively.

The following theorem is valid: *a graph is bipartite if, and only if, there is no odd-membered cycle subgraph of the graph*. Therefore, there is no difficulty in deciding by inspection whether one graph is bipartite or not, and an effective colouring process is unnecessary (see Fig. 3).

All graphs of chemical interest can be coloured in three colours and are therefore tripartite graphs. However this fact has so far given few useful consequences [51].

Hydrocarbons which can be represented by bipartite graphs are called *alternant hydrocarbons* (AH). The concept of AH's was first introduced by Coulson and Longuet-Higgins [52]. Hydrocarbons which can be represented by non-bipartite graphs are called *non-alternant hydrocarbons* (NAH). We use both terms (AH = bipartite graph, NAH = non-bipartite graph) which are, in fact, the graph theoretical and chemical expressions of the same concept.

If one numbers a bipartite graph so that $1, 2, \ldots, p$ are starred and $p+1, p+2, \ldots, p+q = N$ are unstarred vertices, it is obvious that

$$A_{rs} = 0 \text{ for } 1 \leqslant r, s \leqslant p \text{ and } p + 1 \leqslant r, s \leqslant p + q \qquad (30)$$

because two starred or unstarred vertices are never in relation $\mathscr{A}$. Therefore, A is of the form

$$A = \begin{bmatrix} O & B \\ B^+ & O \end{bmatrix} \qquad (31)$$

Eq. (31) was first used by Ham [53]. As an example of matrix A in the form (31) we can give matrix A_2 (Eq. (16)).

III. Equivalence between Elementary Molecular Orbital Theory and the Graph Spectral Problem

A. Hückel Theory

The fundamental fact which makes it possible to apply Graph Theory in quantum chemistry is that the molecular Hamiltonian for a wide class of compounds can be written in the matrix form as a unique function of the graph corresponding to the molecule under consideration, *i.e.*

$$H = H(A). \qquad (32)$$

A linear dependence is the simplest form of Eq. (32),

$$H = \alpha I + \beta A. \qquad (33)$$

Eq. (33) is rather simple, but it can be justified using rigorous quantum mechanical analysis [23]. This type of Hamiltonian was first introduced by Hückel [24,35,54] in the early years of quantum chemistry before Graph Theory was developed. The whole approach is today known as

Hückel theory and it was introduced in order to describe the π electrons in planar conjugated molecules. α and β are the Coulomb and resonance integrals, respectively, of some effective one-electron Hamiltonian operator [55].

From Eq. (33) it can be seen that H and A commute,

$$HA = AH \qquad (34)$$

and have therefore the same eigenvectors. Thus, the *eigenvectors of the adjacency matrix will be identical with the simple Hückel MOs* if the overlap matrix S is assumed to be of the form [23]

$$S = I + \sigma A \qquad (35)$$

where σ represents overlap integral. This can be proved in a simple way. Let C_i be an eigenvector of the matrix A belonging to the eigenvalue x_i, *i.e.*

$$C_i A = x_i C_i. \qquad (36)$$

Then

$$C_i (H - E_i S) = (\alpha + x_i \beta - E_i - x_i E_i \sigma) C_i = O \qquad (37)$$

for

$$E_i = \frac{\alpha + x_i \beta}{1 + \sigma x_i} \qquad (38)$$

σ is usually assumed to be zero (zero-overlap approximation). Using β as an energy unit and α as the zero-energy reference point, we have

$$E_i = x_i \qquad (39)$$

and hence, the *eigenvalues of the adjacency matrix are identical with orbital energy levels.*

Therefore, it is clear that the spectrum of the graph is rather important in elementary MO calculations. The elementary MO problem, *i.e.* Hückel problem, is in fact fully equivalent to the graph spectral problem. This was first emphasized by Günthard and Primas [56] and later by Schmidt-ke [57].

Simple molecular orbitals (eigenvectors) corresponding to $x > 0$, $x < 0$, and $x = 0$ are called bonding, antibonding, and non-bonding, respectively. Note that the number of linearly independent non-bonding molecular orbitals (NBMOs) is equal to the multiplicity of the number zero in the graph spectrum. This fact can be related to the stability of the particular conjugated molecule, and it will be discussed later (see Section VI).

The total π-electron energy of the ground state is given by

$$E_\pi = \sum_{j=1}^{N} g_j \, x_j \tag{40}$$

where g_j is the orbital occupancy number, and for the majority of conjugated molecules:

$$g_j = \begin{cases} 2 \text{ for } x_i > 0 \\ 1 \text{ for } x_i = 0 \\ 0 \text{ for } x_i < 0 \end{cases} \tag{41}$$

We would also like to mention here another important point. Since the Hückel and topological matrices are closely related for a particular conjugated molecule, all properties of a molecule (*i.e.* energy, MOs, bond orders, charge densities) which may be derived from the topological matrix by mathematical treatment must be dependent on the molecular topology [23]. This may be one reason why the predictive power of elementary Hückel theory is in many cases (*e.g.* for alternant hydrocarbons) as good as that of any more elaborate approach [58].

B. The Sachs Theorem

As was shown earlier, the adjacency matrix of a graph (*i.e.* topological matrix of a molecule) is not a unique function of the graph but also depends on the numbering of the vertices. Therefore, it is very important that the characteristic polynomial and hence the spectrum should be independent of the numbering of the vertices.

However, there is not a one-to-one correspondence between a graph and its characteristic polynomial, and this has been the subject of investigation of several authors [59-66]. Non-identical graphs can possess the same spectrum. These are so called *isospectral* graphs. The simplest examples of such graphs interesting for chemists are

G_1 corresponds to 1,4-divinyl benzene, while G_2 corresponds to 2-phenyl butadiene.

The method of calculating the characteristic polynomial by expansion of the determinant is neither simple nor does it give any insight into the relation between the graph structure and the value of the coefficients. This is a fundamental problem in graph spectral theory and it had been the object of numerous investigations [67–71]. The final solution was given ten years ago by Sachs [27]. Here we may point out that Professor Coulson obtained the same theorem years ago [72] and used it in his own work as a further way of expanding secular determinants. Some of his results were published twenty years ago in Ref. [29].

In this section we shall report only the special case of the Sachs theorem which covers the chemically important graphs. To further understanding of the Sachs theorem, the term *Sachs graph* of the graph G has been introduced [73]. A Sachs graph is a subgraph of G which has no components other than complete graphs of degree one and cycles. Thus, for example, G_2, G_3, G_4, G_5, and G_6 are Sachs graphs of the graph G_1.

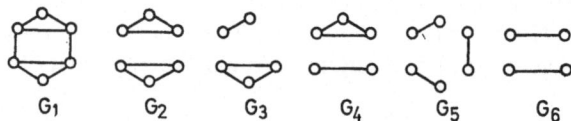

Note that G_4 is different from G_3. The number of components of a Sachs graph s is denoted by $c(s)$ and the number of cycle components (rings) is denoted by $r(s)$. The Sachs graph G_3 has $c(G_3) = 2$ and $r(G_3) = 1$. The coefficients of the characteristic polynomial of the graph (see Eq. (26)) can be obtained by means of the following expression:

$$a_0 = 1$$

$$a_n = \sum_{s \in S_n} (-)^{c(s)} \, 2^{r(s)} \quad 1 \leqslant n \leqslant N \tag{42}$$

Here S_n is the set of all Sachs graphs of the graph G with n vertices and, hence, the summation goes over all Sachs graphs with n vertices. In the case when the set S_n is an empty set ($\emptyset$), there is no Sachs graph with n vertices, and hence $a_n = 0$.

The use of the Sachs formula is illustrated below.

1)

$$S_1 = \emptyset$$

$$S_2 = \left\{ \circ\!\!-\!\!\circ \right\}$$

therefore,

$$a_1 = 0$$
$$a_2 = (-)^1 \, 2^0 = -1,$$

and

$$P_G(x) = x^2 - 1.$$

2)

Therefore, $a_1 = 0$

$$a_2 = (-)^1 \, 2^0 + (-)^1 2^0 + (-)^1 2^0 + (-)^1 2^0 = -4$$
$$a_3 = (-)^1 \, 2^1 = -2$$
$$a_4 = (-)^2 \, 2^0 = 1,$$

and

$$P_G(x) = x^4 - 4x^2 - 2x + 1.$$

The reader can easily see that the Sachs theorem becomes a cumbersome method for enumeration of the coefficients a_n for large molecules. Some other procedures for the enumeration of a_n have also been proposed [29,74, 75]. Hosoya [75] developed his method independently of Sachs; however, it can be shown that the two approaches are fully equivalent, Hosoya's method being more practical [76].

Regardless of the difficulty mentioned, the Sachs theorem has turned out to be a useful tool for studying the spectrum of the graph. Some simple and obvious consequences of Eq. (42) are:

a) Since there is no Sachs graph with only one vertex, it is always $S_1 = \emptyset$ and therefore $a_1 = 0$. This means that the sum of the whole spectrum is zero, *i.e.*

$$\sum_{i=1}^{N} x_i = 0 \qquad (43)$$

b) Any two vertices joined by an edge can represent an element of S_2. Therefore, the number of Sachs graphs with two vertices is equal to the number of edges (v) and hence

$$a_2 = -v \tag{44}$$

c) Similarly

$$a_3 = -2\,n_3 \tag{45}$$

where n_3 is the number of 3-membered cycles in the graph.

The generalization of these considerations for an arbitrary a_n is far from being simple [29]. However, the following general rule is valid for bipartite graphs:

d) Sachs graphs with an odd number of vertices necessarily contain odd-membered cycles. Since there are no odd-membered cycles in bipartite graphs, there is always fulfilled

$$S_n = \phi \text{ and } a_n = 0 \text{ for odd } n. \tag{46}$$

IV. The Pairing Theorem

In Section II the notion of alternant and non-alternant hydrocarbons was introduced, using a graph colouring approach. Although this idea looks interesting, it is rather difficult to see whether there is any physical concept behind the colouring process. In the present section we show the extraordinary importance of separating AHs from NAHs, and in subsequent sections a number of applications will be discussed.

The characteristic polynomial of the bipartite graph according to Eq. (46) is of the form:

$$P_G(x) = x^N + a_2\,x^{N-2} + a_4\,x^{N-4} + \cdots \tag{47}$$

Hence

$$P_G(-x) = P_G(x) \text{ for } N \text{ even} \tag{48}$$
$$P_G(-x) = -P_G(x) \text{ for } N \text{ odd}$$

In both cases, *if x_i is a root of the secular equation $P_G(x) = 0$, $-x_i$ is a root too*. There is at least one zero in the spectrum for odd N. Thus in order to construct the whole set of MOs it is sufficient to find only the positive (or negative) eigenvalues, and this can be used for simplification of the MO calculations [77].

The above result, usually called the *Pairing Theorem*, was first proved in 1940 by Coulson and Rushbrooke [20]. There are a number of proofs and detailed discussions of the pairing theorem in the literature [78]. The pairing theorem claims that the spectrum of a bipartite graph is symmetric regarding $x = 0$. It is interesting that this theorem has only recently been proved by the graph theoretical approach [79,80]. Note that only bipartite graphs have a symmetric spectrum [51,79].

The secular equation for AHs can be written as

$$x_i \overset{*}{c_{ir}} + \sum_{s \to r} \overset{0}{c_{is}} = 0 \qquad \text{for } r = \text{starred vertex}$$

$$x_i \overset{0}{c_{ir}} + \sum_{s \to r} \overset{*}{c_{is}} = 0 \qquad \text{for } r = \text{unstarred vertex}$$

$$(49)$$

where the summation goes over all first neighbours of the vertex r. $\overset{*}{c_{ir}}$ and $\overset{0}{c_{is}}$ are the components of the eigenvector C_i corresponding to starred and unstarred vertices, respectively, that is

$$C_i = (\overset{*}{c_{i1}}, \overset{*}{c_{i2}}, \ldots, \overset{*}{c_{ip}}, \overset{0}{c_{i\,p+1}}, \overset{0}{c_{i\,p+2}}, \ldots, \overset{0}{c_{i\,p+q}}) \qquad (50)$$

where vertices $1, 2, \ldots, p$ are starred and $p + 1, p + 2, \ldots, p + q$ are unstarred, respectively. Since $-x_i$ is also the solution of the secular equation, it is easy to show that

$$C_i^{\text{pair}} = (\overset{*}{c_{i1}}, \overset{*}{c_{i2}}, \ldots \overset{*}{c_{ip}}, -\overset{0}{c_{i\,p+1}}, -\overset{0}{c_{i\,p+2}}, \ldots -\overset{0}{c_{i\,p+q}}) \qquad (51)$$

is an eigenvector with the eigenvalue $-x_i$. Specially for $x_i = 0$

$$\overset{0}{c_{is}} = -\overset{0}{c_{is}} \qquad (52)$$

and the NBMOs are of the form

$$C_i^{\text{NBMO}} = (\overset{*}{c_{i1}}, \overset{*}{c_{i2}}, \ldots, \overset{*}{c_{ip}}, 0, 0, \ldots, 0). \qquad (53)$$

In matrix notation Eqs. (50) and (51) can be represented as

$$C = \begin{bmatrix} C\text{bond.} \\ C\text{antibond.} \end{bmatrix} = \begin{bmatrix} U & V \\ U & -V \end{bmatrix} \qquad (54)$$

$C_{\text{bond.}}$ and $C_{\text{antibond.}}$ are the bonding and antibonding MOs occupied in the ground state by 2 and 0 electrons, respectively. For the sake of simplicity, NBMOs are not considered here. Since the adjacency matrix can be by appropriate numbering of vertices written in the form (31), the eigenvalue matrix X is given as

$$X = CAC^+ = \begin{bmatrix} Y & O \\ O & -Y \end{bmatrix} \tag{55}$$

where

$$UBV^+ = VB^+ U^+ = 1/2Y. \tag{56}$$

Y is a diagonal matrix. This is the matrix formulation of the spectral symmetry.

The colouring process makes explicit the difference between starred and unstarred vertices. Therefore, one is justified in assuming that the starred and unstarred atoms have different properties. One can generalize Eq. (33) as:

$$\tilde{H} = \begin{bmatrix} \alpha^* I & O \\ O & \alpha^\circ I \end{bmatrix} + \beta \begin{bmatrix} O & B \\ B^+ & O \end{bmatrix} \tag{57}$$

where $\alpha^* \neq \alpha^\circ$. It has been shown [81–84] that the solutions of Eq. (57) are related to those of Eq. (33), hence

$$\tilde{E}_i = \frac{\alpha^* + \alpha^\circ}{2} \pm 1/2 \sqrt{(\alpha^* - \alpha^\circ)^2 + 4\beta^2 x_i^2} \tag{58}$$

and

$$\tilde{C}_i = \frac{2 x_i \beta}{\alpha^* - \alpha^\circ \pm \sqrt{(\alpha^* - \alpha^\circ)^2 + 4\beta^2 x_i^2}} C_i \tag{59}$$

$i = 1, 2, \ldots, N/2$

Eqs. (58) and (59) enable, for example, the study of aza boron compounds, as:

V. Bond Orders and Charges

Coulson's charge-bond order matrix [37,52,85,86] is defined as

$$P^C = 2 C^+_{bond.} C_{bond.} \tag{60}$$

Using the following relations:

$$C_{bond.} = (U, V) \text{ and } C^+C = I \tag{61}$$

we obtain

$$PC = \begin{bmatrix} I & 2\ U^+V \\ 2\ V^+U & I \end{bmatrix} \qquad (62)$$

Now, it can be seen that the *π-electron charge on the arbitrary atom r of AHs is equal to unity.*

$$q_r = (P^C)_{rr} = 1. \qquad (63)$$

Similarly, the bond orders between the atoms of the same colour are zero:

$$(P^C)_{**} = (P^C)_{oo} = 0. \qquad (64)$$

A proper consequence of Eq. (63) is the prediction of *zero π component of the dipole moment of AHs.* This prediction can, of course, be checked by measurements of molecular dipole moments. Available experimental data confirm this prediction [87]. Some examples are given in Fig. 4. The

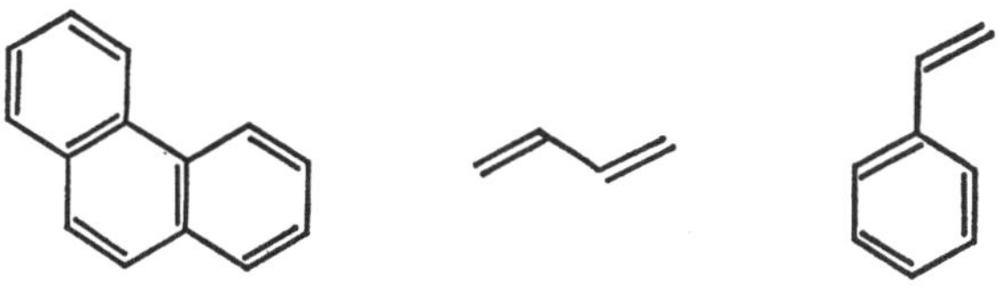

Predicted: 0	0	0
Experimental value : 0^{88}	0^{89}	$0.13^{90,91}$

Fig. 4. Dipole moments of some alternant hydrocarbons

fact that the charge distribution in AHs is uniform is related to the self-consistent nature of topological MOs of AHs [58]. Therefore, elementary MO theory should give good agreement with experimental values, *e.g.* heats of atomization, bond lengths, resonance energies, etc. for AHs. This has recently been found [92,93]. It has been shown [94,95] that self-consistency depends on the exactness of the equations

$$(P^C)_{11} = (P^C)_{22} = \ldots = (P^C)_{NN} = 1. \qquad (65)$$

Elementary MOs are not satisfactory for NAHs in many cases. For example, a number of cases can be found where the stable singlet ground state and relatively large total π-electron energy are obtained from the calculations, but the molecules cannot be made. Some examples are reported below:

The appearance of zero bond orders between the iso-coloured vertices has been used in the perturbation theory approach for the calculation of the total π-electron energy [48,96,97]. For the transformation

$$\tag{66}$$

that is, for the introduction of a new edge in the graph of the molecule, the energy change (δE) in the first order approximation is

$$\delta E = 2\, p_{rs}. \tag{67}$$

Eq. (64) can now be interpreted as follows: the introduction of a bond between the iso-coloured atoms, $i.e.$ the closure of an odd-membered cycle, causes only a small change of second- and higher-order perturbation terms in the electronic energy of the conjugated hydrocarbon. Thus, for example, the 9—10 bond in azulene is rather long ($\sim$1.50 A) [98,99]. Similarly, since the AH (I) is unstable (see Section VI), the molecule (II) should be unstable too, because it cannot gain any stabilization through linking atoms r and s.

I II

(No Kekulé structure)

This is in agreement with the fact that II has not as yet been isolated [100].

From inspection of Eq. (61) it is clear that, the only non-vanishing bond orders in AHs are those between differently coloured vertices, $i.e.$

$$P*^\circ = 2\ U^+V. \tag{68}$$

Hall [101] has shown that

$$P*^\circ = (B^+B)^{-\frac{1}{2}}\ B. \tag{69}$$

Eq. (69) besides giving an excellent possibility for enumerating $P*^\circ$ shows that bond orders are a function of molecular topology only. Since bond orders can be correlated with bond lengths [85,102–104], *molecular topology also determines molecular geometry.*

Coulson's [85] definition of the bond order P^C:

$$(P^C)_{rs} = \sum_{j=1}^{N} g_j c_{jr} c_{js} \tag{70}$$

is not the only one possible. Mulliken [105,106] proposed the following relation

$$(P^M)_{rs} = (1 + \sigma) \sum_{j=1}^{N} g_j\ \frac{c_{jr} c_{js}}{1 + x_j \sigma}. \tag{71}$$

Similarly, Ruedenberg [21] also proposed a simple expression for calculating bond orders

$$(P^R)_{rs} = \sum_{j=1}^{N} g_j c_{jr} c_{js}/x_j. \tag{72}$$

If there is no zero in the spectrum of the graph, Eqs. (70) and (58) are equivalent. The generalization of Eqs. (70)–(72) is

$$(P)_{rs} = \sum_{j=1}^{N} g_j c_{jr} c_{js} f(x_j) \tag{73}$$

and it has been shown [53,107] that

$$P^C = I + \sqrt{A^2}\ A^{-1} \tag{74}$$

$$P = P^C\ f(A). \tag{75}$$

71

Therefore, Mulliken's and Ruedenberg's expressions for the evaluation of bond orders are related to that of Coulson and this should be so because all these expressions are derived in consideration of the molecular topology.

Hall's formula (69) can be obtained from Eq. (74), because for AHs the topological matrix is of the form (31). In the case of NAHs $\sqrt{A^2}\,A^{-1}$ have non-zero diagonal elements; π-electron distribution is thus not uniform and the dipole moments have values very different from zero. The rules governing the charge displacements in NAHs are discussed in Section VII.

Ruedenberg's bond orders (P^R) are identical with Pauling's VB bond orders [53,107]. This is not the only relation between VB and simple MO theory. There is a relation between MOs and Kekulé structures and it will be discussed in detail later on (see Section VIII).

VI. Unstable π-Electron Systems

One of the most intriguing successes of simple MO theory is the prediction of nonexistence of a number of π-electron systems. Chemical existence is a very difficult notion which includes thermodynamic and kinetic requirements [108], so that predictions obtained from MO theory must be interpreted rather cautiously. Since it is assumed that π-electron systems are planar (or nearly so), any considerable deviation from planarity can alter the theoretical predictions. For example, cyclooctatetraene (which is predicted to be a rather unstable compound) and its derivatives are well defined compounds, but have nonplanar (a puckered D_{2d} "tub") conformation [109,110]; it is this escape from planarity which makes cyclooctatetraene a stable compound. Other compounds, may exist as highly reactive species, as is the case with cyclobutadiene [111,112] and its derivatives [113].

There is no a priori chemical reason for any system of an even number of electrons to be nonexistent or highly reactive. The MO considerations show that the instability of certain π-electron systems has a purely topological background. Our discussion will include only AH with an even number of atoms (and π electrons). For NAH, there are some difficulties which so far remain unsolved [114].

For AHs, as shown above, the pairing theorem holds. Suppose first that there is no zero in the spectrum of the bipartite graph. The spectrum possesses, therefore $N/2$ positive and $N/2$ negative eigenvalues and the same is true of energy levels. All the $N/2$ bonding MOs are doubly occupied in the ground state and, hence, a singlet ground state is obtained for the molecule considered.

As a simple consequence of the Pairing theorem there must be even number of zeros in the graph spectrum. Suppose now that there are two zeros in the spectrum, that is, two NBMOs. Thus there are $N/2-1$ bonding and $N/2-1$ antibonding MOs. Since the bonding MOs are doubly occupied, two electrons remain in the two NBMOs and, according to Hund's rule, a triplet (biradical) ground state is expected. This simple consideration was first stated by Longuet-Higgins [115]. The above discussion is illustrated in Fig. 5. Biradicals are very unstable species.

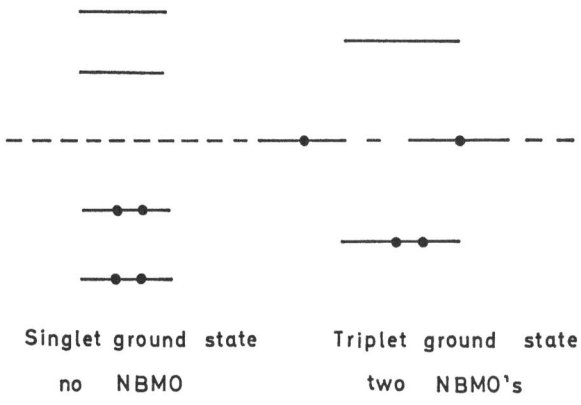

Singlet ground state   Triplet ground state

no NBMO   two NBMO's

Fig. 5.

They are in an orbitally degenerate state and are forbidden according to the Jahn-Teller theorem [116]. Therefore such species either do not exist or are distorted in order to relieve the degeneracy. Good examples of this rule are cyclobutadiene, which has a rectangular shape in the ground state instead of a regular square form [112,117] and cyclooctatetraene, which has a "tub" shape in the ground state instead of, a regular planar octagon [109,110,118,119].

From the above discussion it is obvious that the problem of the instability of alternant π-electron systems is reduced to the finding of necessary and sufficient conditions for the existence of the number zero in the graph spectrum [120]. This problem has not been solved for the general case, but a set of useful theorems concerning the number zero in spectrum is known. We will denote the number of zeros in the graph spectrum as η.

Theorem 1 [114]

Let there be p starred and q unstarred atoms in the AH. Then

$$\eta \geqslant p-q. \tag{76}$$

If there are no $4m$-membered cycle subgraphs of the graph,

$$\eta = N - 2t \tag{77}$$

is valid, where t is the largest possible number of double bonds which can be written in the structural formula of a particular conjugated molecule. The existence of even one single Kekulé structure (*i.e.* $N = 2t$) means that the molecule does not have any zeros in the spectrum, and consequently it should be stable. This theorem holds for a number of conjugated molecules: polyacenes, polyenes, etc. It has been proved in Ref. [114] using the Sachs theorem. Close to this is Dewar's theorem no. 27 (see Ref. [96]).

It can be seen that, if $p \neq q$, no Kekulé structure can be written for a molecule. Therefore, an important chemical consequence of Theorem 1 is that compounds without Kekulé structures are biradicals. This empirical rule [121] was first deduced from simple MO considerations by Dewar and Longuet-Higgins [115,122].

Theorem 2 [114,120)]
If there is a vertex of degree one, we can remove it and its neighbour and all adjacent edges without changing the value of η:

Example:

Theorem 3 [114)]
A chain of four vertices can be replaced by an edge without changing the value of η:

Example:

We emphasize here that in this theorem the number four has a fundamental meaning. Many of molecular properties are of the modulo four as
will be shown later.

Theorem 4 [114]

Two vertices and four edges of a peripheral 4-membered ring can be
removed without changing the value of η:

$$\eta\left(\boxed{}\right) = \eta\left(\boxed{}\right)$$

Example:

$$\eta\left(\text{⬡⬡}\right) = \eta\left(\text{⬡}\right) = \eta\left(\text{⬡}\right) = \eta\left(\vcenter{\hbox{$\vdots$}}\right) = 2$$

Theorem 5 [114,120]

Let the graph G be of the form

$$\text{(G}_1\text{)} \underset{G}{\rule{2cm}{0.4pt}} \text{(G}_2\text{)}$$

and $\eta\,(G_1) = 0$. Then $\eta\,(G) = \eta\,(G_2)$.

Example:

$$\eta\left(\text{⬡◇⬡}\right) = \eta\left(\text{◇⬡}\right) = \eta\left(\text{◇}\right) = 2$$

The importance of this theorem for chemistry is that the introduction of
vinyl, phenyl, and similar groups at arbitrary positions in the molecule
cannot cause either essential stabilization of an otherwise unstable
molecule or destabilization of an otherwise stable molecule.
For example

I   II

75

Molecule (I) is stable (singlet ground state), while molecule (II) is unstable (triplet ground state). Cyclobutadienes are unstable regardless of the substituents. There are a number of experimental supports for this general conclusion [113,123].

Theorem 6 [124]
If the graph shows a detail

then $\eta > 0$.

Theorem 7 [115,122]
AH's having at least one Kekulé structure and without $4m$-membered cycles have no zero in the spectrum. Moreover, Dewar and Longuet-Higgins [115,122] have given the necessary and sufficient condition for the existence of zeros in the spectrum of AH's possessing $4m$-membered cycles. This was also discussed by Wilcox [125] and Graovac *et al.* [73]. Since the Dewar-Longuet-Higgins rules demand a knowledge of all Kekulé structures of a molecule, their practical value is limited. Nevertheless, they show how the ideas of Resonance Theory play a significant role in MO theory.

The problem of estimating the number of zeros in the graph spectrum has also been investigated by various other authors [126–130].

Chemical stability has a thermodynamic and kinetic background[108] and the energy difference between the highest occupied MO (HOMO) and lowest unoccupied MO (LUMO) reflects the ability of a molecule to react [131]. Therefore, a small HOMO-LUMO separation is indicative of high reactivity. The prediction is made that the lower value of the ionization potential should correlate with the higher reactivity (and instability) of a conjugated molecule [132–134]. There are some experimental data now available [135,136] which show that the reactivity of a series of molecules can be correlated with their ionization potentials. It was pointed out some time ago that there is a close connection between the excitation energies of conjugated systems and their aromaticity [137].

Since all occupied energy levels fall between 0 and 3 (in β units), HOMO-LUMO separation decreases with the increase in molecular size. This fact is reflected in the chemical behaviour of polyenes [138] and polyacenes [139]. Similarly, conjugated macromolecules are found to be paramagnetic [140].

This discussion shows the importance of the zeros in the graph spectrum. Since graph spectra are related only to elementary MO theory, the meaning of the zero in the spectrum is that the *exact* HOMO-LUMO separation is rather small. Then a singlet ground state is expected and a very small excitation energy is needed for the promotion of the electron to LUMO level (thermal excitation). Hence, the same conclusion can be reached about the high chemical reactivity. This is confirmed by the chemistry of a very reactive molecule, cyclobutadiene. When cyclobutadiene is studied with the SCF–MO approach [112], it is predicted to have a rectangular singlet ground state. On the other hand, there is evidence [113] that a derivative of cyclobutadiene, tetramethylcyclobutadiene, shows radical behaviour in the gas phase. This may be a case when a low-lying excited state is populated by the thermal excitation.

VII. The Hückel ($4m + 2$) Rule and its Generalizations

The approximate MO study of annulenes is a problem which can be solved in closed analytical form. The graph representation of N-annulene is a N-membered cycle, and therefore the use of the D_{Nh} symmetry group is allowed. The graph spectrum and eigenvectors are known from the classical work of Hückel [35]. The explicit expression for the total π-electron energy of annulene was obtained later [40,41]. Hence, the energy expression of N-annulene can be evaluated using the following expressions

$$\pi \text{ (annulene)} = \begin{cases} 4 \cosec \dfrac{\pi}{N} & \text{for } N = 4m + 2 \\ 4 \cot \dfrac{\pi}{N} & \text{for } N = 4m \end{cases} \tag{78}$$

The individual orbital energy can be obtained from

$$E_j = 2 \cos \frac{2j\pi}{N} \qquad j = 1, 2, \ldots, N \tag{79}$$

The rule, which is due to Hückel [35], states that only annulenes of ($4m + 2$)-type are stable and those of ($4m$)-type are not.

The major importance of the Hückel rule is that its predictions can be experimentally checked and hence the predictive power of the Hückel MO theory (*i.e.* topological MO theory) can be critically evaluated. The efforts made to prove or disprove the Hückel rule strongly influenced the development of annulene chemistry [141,142].

A. Hückel Rule

In the last section it was shown that the existence of zeros in the graph spectrum is a sufficient condition for the biradical nature of the molecule. Using Theorem 3, all even cycles can be reduced to 4- or 6-membered ring systems. Calculations show that there is no zero in the spectrum of a 6-membered cycle (*i.e.* benzene) and that there are two zeros in the spectrum of a 4-membered cycle (*i.e.* cyclobutadiene). Another approach to this problem is given by Rouvray [130]. We can also make use of the Sachs Theorem [27]. The coefficient a_N of the characteristic polynomial is given as:

$$a_N = x_1 x_2 \ldots x_N = \prod_{i=1}^{N} x_i. \tag{80}$$

If there are zeros in the graph spectrum, it is

$$a_N = 0. \tag{81}$$

A cycle with N vertices has three Sachs graphs with N vertices, *i.e.*:

and thus

$$a_N = (-)^{\frac{N}{2}} 2^0 + (-)^{\frac{N}{2}} 2^0 + (-)^1 2^1 = \begin{cases} 0 \text{ for } N = 4m \\ -4 \text{ for } N = 4m + 2 \end{cases}$$

The Hückel rule for annulenes is generally in agreement with experimental evidence. Another point is worth mentioning here. Steric effects are also important in annulene series. In some cases the repulsion between the annulene inner hydrogen atoms might be considerable. Therefore the molecule, in order to avoid the effect of hydrogen interference, might distort from planarity. In fact, there is some experimental evidence [109, 143-145] available that up to [18]-annulene the nonplanarity of the molecule is significant.

Of course, the simple considerations presented here cannot take into account the stereochemical features of the molecule and thus all predictions concern only idealized planar annulenes.

Recently it has been shown[93] that the Hückel π-electron energy can be correlated with the total $(\sigma + \pi)$ energy of the molecule and hence with the measurable thermodynamic quantities (heat of atomization). The importance of knowing the total energy is thus clear. The Hückel rule states that $(4m)$-annulenes are poorer in π-electron energy than $(4m + 2)$-annulenes. This fact can be demonstrated, as shown in Fig. 6, by considering the difference in the total π-electron energy between the annulene and the corresponding linear polyene (this difference is called by some authors the Dewar resonance energy [92,93] and by others the index of aromatic stabilization [133,134,146]). Note that larger

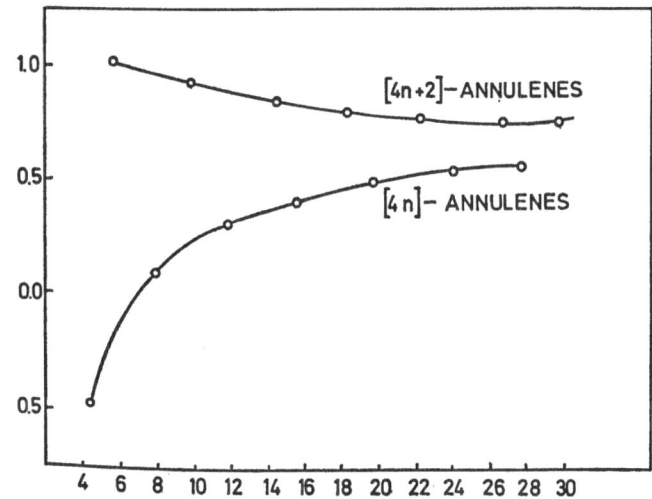

Fig. 6. Index of aromatic stabilization (A_s) vs. the annulene ring size (N)

rings have a smaller effect on the π-electron energy and the Hückel rule is no longer valid when N is sufficiently large. This is in agreement with the prediction of Longuet-Higgins and Salem [147]. They predicted that when the ring becomes very large the Hückel rule will no longer operate. For example, the preliminary studies of the n.m.r. of two derivatives of [30]-annulene (pentadehydro-, tridehydro-) indicated [148,149] that these compounds are cyclopolyenes. However, these results are only tentative and further work may clear up this question. More accurate calculations [150] have given the same result.

Using the perturbation theory approach, it can be shown [39,48,98,151, 152] that the Hückel rule is valid for an arbitrary ring closure of the type:

Such transformations are accompanied by a gain in π-electron energy if $(4m + 2)$-membered rings are formed, and a loss if $(4m)$-membered rings are formed. The closure of odd-membered rings has no energy effect in the first approximation [153].

These results are generally in agreement with chemical experience, since 4-, 8-, and similar rings are rather unusual in conjugated molecules, and their occurrence in the molecule is very often accompanied by poor chemical stability; on the other hand, the 6-membered ring systems are the commonest conjugated molecules.

There is another possible approach, due to Goldstein and Hoffmann [154]. They have shown that the Hückel rule is in agreement with the behaviour of the HOMOs. In the Goldstein -Hoffmann method one considers only the signs of the HOMO coefficients of chains (linear polyenes) and investigates an imaginary ring closure transformation by introducing a new edge into the graph of the linear polyene. If the new edge is introduced between two vertices of the same sign, the ring has aromatic properties, otherwise it is antiaromatic. The coefficient signs of butadiene and hexatriene HOMOs are given below

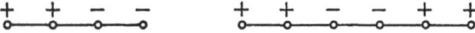

Now we can easily see that the joining of the terminal vertices of the butadiene graph is unfavourable, while the same transformation in the hexatriene graph is favourable.

This is a natural generalization of the Woodward-Hoffmann (W-H) rules [155,156]. A similar approach was also discussed by Salem [157]. The extraordinary success of the W-H rules shows that in some cases purely topological factors govern the course of chemical reactions [158]. An interesting approach was also developed by Dewar and advocated by him for twenty years [159]. He established the following set of rules, now known as the Dewar-Evans (D-E) rules [160]:

a) Thermal pericyclic reactions take place preferentially via aromatic transition states;

b) Photochemical pericyclic reactions lead to products that are formed via antiaromatic transition states;

c) Transition metals may catalyze pericyclic reactions if, and only if, they involve antiaromatic transition states.

Close inspection of the D-E rules shows that the transition states in pericyclic reactions resemble structures which are topologically equivalent to benzene ((4m + 2)ring) or cyclobutadiene ((4m)ring). This only shows that the *basic principles of both the W–H and D–E rules are based on the molecular topology, and thus predictions from both approaches should lead to the same results,* as is indeed the case. Therefore the principles governing the pericyclic reactions must be topology-dependent and *are based on the Hückel rule.*

One of the most famous results of the simple MO approach is the prediction [24] of stability for the cyclopentadienyl anion and cycloheptatrienylium (tropylium) cation, which has been fully confirmed by the experimental findings [161]. The π-electron configuration of these ions is fairly similar to that of benzene, and this explains the results obtained in the first approximation.

It is less clear why the (4m + 1)-membered rings tend to be negatively charged, while the (4m + 3)-membered rings exhibit the tendency to become positively charged in all NAHs. This is generally true, and it can be considered as an extension of the Hückel rule for NAHs [49]. The Hückel rule for NAHs was also discussed in Refs. [162–165]. Graph theory has as yet no explanation for this interesting phenomenon [166].

The Hückel rule can be used in a simple manner for predicting the direction of the dipole moments of NAH as shown in Fig. 7.

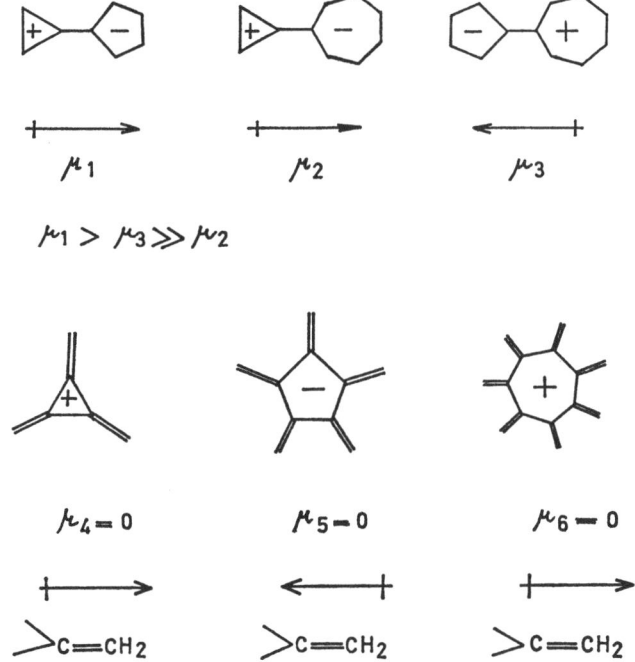

Fig. 7. The direction of dipole moments in some non-alternant hydrocarbons

B. Loop Rule

It can be shown [167] that the Hückel rule is only a corollary of a fairly universal principle determining the total π-electron energy of conjugated molecules. If the graph eigenvalues x_j are ordered by decreasing values, it follows that:

$$E_\pi = 2 \sum_{j=1}^{N/2} x_j. \tag{82}$$

Suppose that the all $N/2$ eigenvalues in the summation appearing in Eq. (82) are positive (or zero) and that the other $N/2$ eigenvalues are negative (or zero), *i.e.*

$$x_1 \geqslant x_2 \geqslant \ldots \geqslant x_{N/2} \geqslant 0 \geqslant x_{N/2+1} \geqslant \ldots \geqslant x_N. \tag{83}$$

Condition (83) is automatically fulfilled for AHs because of the Pairing Theorem. Moreover, condition (83) also holds for the majority of NAHs (but, of course, not for all). Since

$$\sum_{j=1}^{N} x_j = 0 \tag{84}$$

and if relation (83) holds, it is simple to prove that the following equation is valid

$$E_\pi = \sum_{j=1}^{N} |x_j|. \tag{85}$$

Starting from an identical equation, that is,

$$E_\pi = \mathrm{Tr}\, |A| \tag{86}$$

where Tr is the trace of the corresponding matrix, *i.e.* the sum of its diagonal elements, Stepanov and Tatevsky [168,169] derived approximate expressions for E_π of benzenoid hydrocarbons, which is a function of the number of certain type of edges only.

Eq. (85) is at least approximately correct for all conjugated hydrocarbons.

We can expand $|x|$ as a polynomial

$$|x| = l_0 + l_2 x^2 + l_4 x^4 + \ldots \tag{87}$$

and therefore

$$E_\pi = l_0 N + l_2 \sum_{j=1}^{N} x_j^2 + l_4 \sum_{j=1}^{N} x_j^4 + \ldots \tag{88}$$

From elementary matrix algebra it is known that

$$\sum_{j=1}^{N} x_j^n = \mathrm{Tr}A^n. \tag{89}$$

As was shown in Section II, $(A^n)_{rs}$ is the number of paths of length n between the vertices r and s. Therefore, $(A^n)_{rr}$ is the number of loops of length n of the vertex r, and hence $\mathrm{Tr}A^n$ is the number of all loops of length n in the graph. We will call this number L_n and thus

$$\mathrm{Tr}A^n = L_n \tag{90}$$

$$E_\pi = l_0 L_0 + l_2 L_2 + l_4 L_4 + \dots \tag{91}$$

Since the coefficients l_i are obtained from the expansion of $|x|$, the only parameter determining the total π-electron energy is the *number of loops*. Therefore, E_π has a purely combinatorial origin in simple MO theory.

Calculations [167] show that

$$l_{4m+2} > 0 \quad m = 0,1,\dots \tag{92}$$

$$l_{4m} < 0 \quad m = 1,2,\dots$$

In other words, loops of the $(4m+2)$-type make a positive contribution to E_π, while $(4m)$ loops decrease its value. In accordance with

$$|l_2| \gg |l_4| \gg |l_6| \gg \dots \tag{93}$$

the shorter loops have a more pronounced influence on the π-electron energy.

It is important to note that the odd loops (which, of course, do not exist in AHs) play a negligible role in NAHs.

The above-mentioned rule, which we would like to call *"the loop rule"*, is the most general formulation of the $(4m+2)$-type rules. The Hückel rule is obviously a consequence of it.

As an important application of the loop rule, approximate formulae can be derived to show the dependence of E_π on several graph parameters. The problem of finding the dependence of E_π on molecular topology has been investigated by a number of authors [101,167–170], Hall's formula [101] being an example:

$$E_\pi = \frac{7}{4}N + \frac{5}{4}\nu - \frac{1}{64}\sum\sigma_1^2 - \frac{1}{64}\sum\sigma_2. \tag{94}$$

It also includes as a special case McClelland's [170] result. Here N and ν are the number of vertices and edges, σ_1 and σ_2 being the number of the first and second neighbours of a vertex; summation goes over all vertices.

Semiempirical formulae are also given in order to correlate E_π with molecular structure (*i.e.* with some graph parameters) to make direct calculations unnecessary [40,96,171–178].

VIII. Molecular Orbitals and Kekulé Structures

The number of unexcited resonance forms of a conjugated unsaturated hydrocarbon, that is the number of Kekulé structures, play a significant role in organic chemistry [179]. In the past there were attempts to evaluate the number of Kekulé structures for benzenoid hydrocarbons [17,18]. Despite the fact that MO theory and Resonance Theory appear to be independent, there is an interesting relationship between them, and the number of Kekulé structures (K) have an important meaning in elementary MO theory. This is not surprising since the VB Hamiltonian for conjugated hydrocarbons is also completely defined by molecular topology. Details can be found in Pauling's classical papers [180–182], and we note that his *"island"* method is in fact a pure graph-theoretical approach. Similarly, the same is also valid for Rumer's theorem [183] determining the number of linearly independent resonance forms.

Longuet-Higgins [115,184] was the first to show that the number K appears in MO theory; following him, a number of authors [53,73,75,107, 122,125,185,186] have made contributions to this problem. Kekulé structures can be represented in Graph Theory by omitting all bonds except carbon-carbon double bonds, as shown for naphthalene in Fig. 8. Obviously, there is a one-to-one correspondence between a Kekulé structure and a Kekulé graph. The reader can easily see that *Kekulé graphs are Sachs graphs with N vertices without cycles*. This is an important fact because it permits the use of the Sachs Theorem.

We can number the starred vertices 1, 2, ..., p and the unstarred with $1'$, $2'$, ..., p', in an arbitrary way. Every Kekulé graph can now be represented with a permutation $(\alpha, \beta, \ldots, \omega)$ of the numbers $1'$, $2'$, $\ldots p'$, where, by definition, the vertices 1 and α, 2 and β, $\ldots$, p and ω

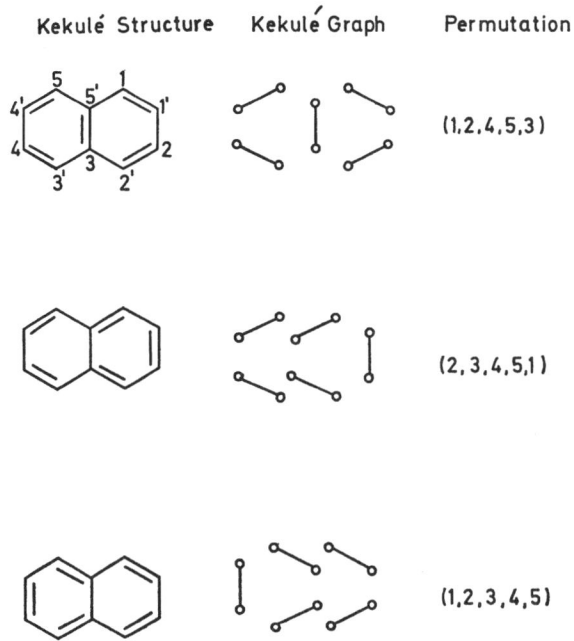

Fig. 8. Kekulé graphs of naphthalene and their permutations

are joined by an edge. There are, of course, K such permutations. These permutations corresponding to the Kekulé graphs for naphthalene are also shown in Fig. 8. The following equation was proved in Refs. [115] and [122]:

$$a_N = (-)^N \det \boldsymbol{A} = (-)^N \left(\sum_{j=1}^{K} p_j\right)^2 \tag{95}$$

where p_j is the parity of the permutation corresponding to the j-th Kekulé graph, $i.e.$

$$p_j = \begin{cases} +1 \text{ if the } j\text{-th permutation is even} \\ -1 \text{ if the } j\text{-th permutation is odd} \end{cases} \tag{96}$$

It can be shown that the parity of all permutations is the same in AHs without $(4m)$-membered rings, and it gives [19]

$$K^2 = \det \boldsymbol{A}. \tag{97}$$

Another form of expression (97) is as follows. The product of all occupied orbital energies (in β units) is equal to the number of Kekulé structures:

$$\prod_{j=1}^{N/2} x_j = K \tag{98}$$

This is an excellent way either to enumerate K or to check the HMO calculations. A detailed graph-theoretical interpretation of Eqs. (95)—(98) is given by Graovac et al. [73].

Another rather interesting relation between the resonance theory and elementary MO theory was established by Ham and Ruedenberg [107]. Let P^P be the Pauling bond-order matrix known from resonance theory [179]. If the adjacency matrix of the j-th Kekulé graph is A_j, then

$$P^P = \frac{1}{K} \sum_{j=1}^{K} A_j. \tag{99}$$

It was shown [53,107] that

$$(P^P)_{rs} = (A^{-1})_{rs} \tag{100}$$

if r and s are starred and unstarred vertices, respectively. The A^{-1} matrix can be obtained by inspection of molecular topology, using only VB ideas, by a method of Heilbronner [185]:

$$(A^{-1})_{rs} = \frac{1}{K} (-)^{\frac{d(r,s)-1}{2}} K_{rs} \tag{101}$$

where K_{rs} is the number of Kekulé graphs after omitting the vertices r and s; d is the distance function (see Section II). Since formula (101) holds only for AHs without $(4m)$-membered rings, there are opinions that for this class of hydrocarbons the elementary MO and VB approaches are fully equivalent [185,186]. Recently, van der Hart, Mulder and Ooysterhoff [156] using the extended VB theory have shown that the same results can be obtained with VB theory as with MO theory.

A generalization of Eq. (97) was proposed [19,187], which shows the dependence of the number of Kekulé structures on molecular topology for arbitrary hydrocarbons. The Sachs theorem is used in order to derive the following equation:

$$K^2 = \operatorname{per} A - \sum_{s \in S_N^{\text{odd}}} 2^{r(s)} \tag{102}$$

where S_N^{odd} is the set of all Sachs graphs with N vertices containing odd-membered cycles. For AHs we have, of course, $S_N^{odd} = \emptyset$, and thus

$$K^2 = \text{per } \boldsymbol{A} \tag{103}$$

per $\boldsymbol{A}$ is the permanent[b] of the matrix $\boldsymbol{A}$.

Furthermore, per $\boldsymbol{A} = \det \boldsymbol{A}$ for graphs which do not contain $(4m)$-membered cycles. The use of Eq. (102) is illustrated on the pyracylene molecule:

$$\text{per } \boldsymbol{A} = 20 \qquad K = \sqrt{20-2^2} = 4$$

IX. Unsolved Problems and Future Developments

In the preceding chapters a number of unsolved problems were discussed. Almost all the theorems mentioned in this paper concern AHs, while NAHs are an open field for investigations by means of Graph Theory. The rules governing charge distributions in NAHs are rather interesting.

Although VB and simple MO theories are obviously closely related (see Section VIII), we are not sure why and for which classes of molecules the results of VB and MO calculations are equivalent.

Conjugated hydrocarbons are not the only type of molecules convenient for application of Graph Theory. There is a priori no reason why

---

[b] The permanent of a $N \times N$ square matrix $\boldsymbol{X} = [x_{ij}]$ is defined as

$$\text{per } \boldsymbol{X} = \sum_P x_{1\alpha}\, x_{2\beta} \ldots x_{N\omega}$$

where $(\alpha, \beta, \ldots, \omega)$ is a permutation P of the indices $1, 2, \ldots, N$ and the summation is over all $N!$ permutations. There is a near relationship between the permanent and the determinant, since

$$\det \boldsymbol{X} = \sum_P (-)^P x_{1\alpha}\, x_{2\beta} \ldots x_{N\omega}$$

There is a detailed discussion of this topic in Ref. [188].

topological MOs should not be able to describe every class of molecules correctly. The best results to date have been obtained for inorganic complexes [189-192] and for boron hydrides [193-195]. Boron compounds seem to be the three-dimensional analogues of conjugated hydrocarbons [193] and a generalization of Hückel-type calculations for the three-dimensional case is under way [195].

As indicated in Section VII, the molecular topology can govern the mechanism of some chemical reactions (W–H and D–E rules). The major importance of HOMOs and LUMOs for both ground and transition states is merely sketched by Goldstein and Hoffmann [154]. The future will, we hope, yield a properly developed topological theory of chemical reactivity.

The mathematical apparatus of Graph Theory is rather simple and therefore very large molecules, even infinite ones, can be investigated [196-200]. It would be of great value if biologically important compounds (*e.g.* steroids, proteins, DNA, etc.) could also be treated in such a simple manner. A very interesting application in this area may be an attempt to study the σ-electronic systems using graph theory [201].

Acknowledgements. We would like to thank Dr. L. Klasinc, Dr. T. Cvitaš, Mr. M. Milun and Mr. A. Graovac for their valuable comments and help in the presentation of this work.

X. References

[1] Avondo-Bodino, G.: Economic applications of the theory of graphs. New York: Gordon and Breach 1962.
[2] Mattuck, R. D.: A guide to Feynman diagrams in the many-body problem. New York: McGraw-Hill 1967.
[3] Harary, F.: Graph theory and theoretical physics. New York: Academic Press 1966.
[4] Cartwright, D., Harary, F.: Psychol. Rev. *63*, 277 (1963).
[5] Lane, R.: Elemente der Graphentheorie und ihre Anwendung in den biologischen Wissenschaften. Leipzig: Akademischer Verlag 1970.
[6] Čulik, K.: Application of graph theory to mathematical logics and linguistics. Prague: Czechoslovak. Acad. Sci. 1964.
[7] Flament, C.: Applications of graph theory to group structure. New York: Prentice-Hall 1963.
[8] Harary, F.: Graph theory. Reading, Mass. Addison-Wesley 1969.
[9] See, for example: Rouvray, D. H.: R. I. C. Rev. *4*, 173 (1971).
[10] Professor J.-E. Dubois (L'Université de Paris) has kindly informed us about a project in connection with the categorization of chemical systems going on in his laboratory.
[11] Cayley, A.: Phil. Mag. *67*, 444 (1874).
[12] Pólya, G.: Acta Math. *68*, 145 (1937).

[13] Harary, F.: Usp. Mat. Nauk (USSR) 24, 179 (1969).
[14] Ruch, E., Hasselbarth, W., Richter, B.: Theoret. Chim. Acta 19, 288 (1970).
[15] Rouvray, D. H.: Chemistry 45, 6 (1972). — Ege, G.: Naturwissenschaften 58, 247 (1971).
[16] Balaban, A. T.: Tetrahedron 27, 6115 (1971), and references therein.
[17] Gordon, M., Davison, W. H. T.: J. Chem. Phys. 20, 428 (1952).
[18] Yen, T. F.: Theoret. Chim. Acta 20, 399 (1971).
[19] Cvetković, D., Gutman, I., Trinajstić, N.: Chem. Phys. Letters 16, 614 (1972).
[20] Coulson, C. A., Rushbrooke, G. S.: Proc. Camb. Phil. Soc. 36, 193 (1940).
[21] Ruedenberg, K.: J. Chem. Phys. 22, 1878 (1954).
[22] Ham, N. S., Ruedenberg, K.: J. Chem. Phys. 29, 1199 (1958).
[23] Ruedenberg, K.: J. Chem. Phys. 34, 1861 (1961).
[24] Hückel, E.: Z. Physik 70, 204 (1931); 72, 310 (1931); etc.
[25] For some basic definitions of Graph Theory see also: Essam, J. W., Fisher, M. E.: Rev. Mod. Phys. 42, 272 (1970).
[26] Berge, C.: The theory of graphs and its applications. London: Methuen 1962.
[27] Sachs, H.: Publ. Math. (Debrecen) 11, 199 (1963).
[28] Balandin, A. A.: Acta Physicochim. U.S.S.R. 12, 447 (1940).
[29] Coulson, C. A.: Proc. Camb. Phil. Soc. 46, 202 (1949).
[30] Gantmaher, F. R.: Matrix Theory (in Russian), Nauka. Moscow 1967.
[31] Cvetković, D.: Publ. Fac. Electrotechnique Univ. Belgrade 354—356, 1 (1971).
[32] Schuster, P.: Theoret. Chim. Acta 3, 278 (1965).
[33] Wild, U., Keller, J., Günthard, H. H.: Theoret. Chim. Acta 14, 383 (1969).
[34] Petersdorf, M., Sachs, H.: Wiss. Z. Techn. Hochsch. Ilmenau (D.D.R.) 15, 123 (1969).
[35] Hückel, E.: Z. Physik 76, 628 (1932).
[36] Coulson, C. A.: Proc. Phys. Soc. (London) 60, 268 (1947).
[37] Coulson, C. A., Longuet-Higgins, H. C.: Proc. Roy. Soc. (London) A 191, 39 (1947); A 193, 447 (1948).
[38] Bradburn, M., Coulson, C. A., Rushbrooke, G. S.: Proc. Roy. Soc. (Edinburgh) A 62, 336 (1948).
[39] Heilbronner, E.: Helv. Chim. Acta 37, 921 (1954).
[40] Peters, D.: J. Chem. Soc. 1958, 1023, 1028, 1031.
[41] Polansky, O. E.: Monatsh. Chem. 91, 916 (1960).
[42] Weltin, E., Gerson, F., Murrell, J. N., Heilbronner, E.: Helv. Chim. Acta 44, 1400 (1961).
[43] Golebiewski, A.: Roczniki Chem. 36, 1511 (1962).
[44] Heilbronner, E.: Tetrahedron Letters 1923 (1964).
[45] Heilbronner, E.: Theoret, Chim. Acta 4, 64 (1966).
[46] Kowalewski, M., Golebiewski, A.: Acta Phys. Polon. 35, 585 (1969).
[47] Gutman, I., Milun, M., Trinajstić, N.: Croat. Chem. Acta 44, 207 (1972).
[48] Gutman, I., Trinajstić, N., Živković, T.: Chem. Phys. Letters 14, 342 (1972).
[49] Gutman, I., Trinajstić, N., Živković, T.: Croat. Chem. Acta 44, 501 (1972), Tetrahedron, in press.
[50] Ore, O.: The four color problems. New York: Academic Press 1967.
[51] Rouvray, D. H.: Compt. Rend. 275C, 657 (1972).
[52] Coulson, C. A., Longuet-Higgins, H. C.: Proc. Roy. Soc. (London) A 192, 16 (1948).
[53] Ham, N. S.: J. Chem. Phys. 29, 1229 (1958).
[54] Hückel, E.: Z. Physik 83, 632 (1933).
[55] See, for example, Salem, L.: The molecular orbital theory of conjugated systems. New York: Benjamin 1966.

56) Günthard, H. H., Primas, H.: Helv. Chim. Acta *51*, 1675 (1956).
57) Schmidtke, H. H.: J. Chem. Phys. *45*, 3920 (1966).
58) Dewar, M. J. S.: The molecular orbital theory of organic chemistry. New York: McGraw Hill 1969.
59) Harary, F.: SIAM Review *4*, 202 (1962).
60) Bruck, R. H.: Pacific J. Math. *13*, 421 (1963).
61) Baker, G. A.: J. Math. Phys. *7*, 2238 (1966).
62) Fisher, M.: J. Combinatorial Theory *1*, 105 (1966).
63) Ponstein, J.: SIAM J. Appl. Math. *14*, 600 (1966).
64) Turner, J.: SIAM J. Appl. Math. *16*, 520 (1968).
65) Doković, D. Ž.: Acta Math. Acad. Sci. Hung. *21*, 104 (1970).
66) Balaban, A. T., Harary, F.: J. Chem. Docum. *11*, 258 (1971).
67) Samuel, I.: Compt. Rend. *229*, 1236 (1949).
68) Gourané, R.: J. Rech. Centre Natl. Rech. Sci. *34*, 81 (1956).
69) Collatz, L., Sinogowitz, U.: Abhandl. Math. Sem. Univ. Hamburg *21*, 63 (1957).
70) Spialter, L.: J. Am. Chem. Soc. *85*, 2012 (1963).
71) Spialter, L.: J. Chem. Docum. *4*, 261, 269 (1964).
72) Coulson, C. A.: private communication, December 1972.
73) Graovac, A., Gutman, I., Trinajstić, N., Živković, T.: Theoret. Chim. Acta (Berlin) *26*, 67 (1972).
74) Heilbronner, E.: Helv. Chim. Acta *36*, 170 (1953).
75) Hosoya, H.: Bull. Chem. Soc. Japan *44*, 2332 (1971); Theoret. Chim. Acta *25*, 215 (1972).
76) Hosoya, H.: private communication, June 1972.
77) Moffit, W.: J. Chem. Phys. *26*, 424 (1957).
78) See, for example, Koutecký, J.: J. Chem. Phys. *44*, 3702 (1966).
79) Cvetković, D.: Matematička Biblioteka (Beograd) *41*, 193 (1969).
80) Rouvray, D. H.: Compt. Rend. *274C*, 1561 (1972).
81) Bochvar, D. A., Stankevich, I. V., Chistyakov, A. L.: Zh. Fiz. Khim. *35*, 55 (1961).
82) Bochvar, D. A., Stankevich, I. V., Chistyakov, A. L.: Zh. Fiz. Khim. *39*, 1365 (1965).
83) Bochvar, D. A., Stankevich, I. V.: Zh. Fiz. Khim. *39*, 2028 (1965).
84) Bochvar, D. A., Stankevich, I. V.: Zh. Fiz. Khim. *40*, 2626 (1966).
85) Coulson, C. A.: Proc. Roy. Soc. (London) *A 169*, 413 (1939).
86) Coulson, C. A., Longuet-Higgins, H. C.: Proc. Roy. Soc. (London) *A 195*, 188 (1948).
87) McClellan, A. L.: Tables of experimental dipole moments. San Francisco: Freeman and Co. 1963.
88) Lumbroso, H.: Compt. Rend. *228*, 1425 (1949); Ann. Fac. Sci. Univ. Toulouse *14*, 108 (1950).
89) Hannay, N. B., Smyth, C. P.: J. Am. Chem. Soc. *65*, 1931 (1943).
90) Petro, A. J., Smyth, C. P.: J. Am. Chem. Soc. *79*, 6142 (1957).
91) Petro, A. J.: J. Am. Chem. Soc. *80*, 73 (1958).
92) Hess, B. A., Schaad, L. J.: J. Am. Chem. Soc. *93*, 305, 2413 (1971).
93) Schaad, L. J., Hess, B. A.: J. Am. Chem. Soc. *94*, 3068 (1972).
94) Kirsanov, B. P., Bazilevski, M. V.: Zh. Strukt. Khim. *5*, 99 (1964).
95) Meschetkin, M. M.: Vest. Leningr. Univ. *4*, 12 (1960).
96) Dewar, M. J. S.: J. Am. Chem. Soc. *74*, 3341, 3345, 3350, 3357 (1952).
97) Dewar, M. J. S.: Tetrahedron *8S*, 75 (1966).
98) Hanson, A. W.: Acta Cryst. *19*, 19 (1965).
99) Bastiansen, O., Derissen, J. L.: Acta Chem. Scand. *20*, 1319 (1966).

100) Baumgartner, P., Weltin, E., Wagnieré, G., Heilbronner, E.: Helv. Chim. Acta *48*, 751 (1965).
101) Hall, G. G.: Proc. Roy. Soc. (London) *A 229*, 251 (1955).
102) Coulson, C. A., Golebiewski, A.: Proc. Phys. Soc. (London) *78*, 1310 (1961).
103) Boyd, G. V., Singer, N.: Tetrahedron *22*, 3383 (1966).
104) Živković, T., Trinajstić, N.: Can. J. Chem. *47*, 697 (1969).
105) Mulliken, R. S.: J. Chim. Phys. *46*, 647 (1949).
106) Mulliken, R. S.: J. Chem. Phys. *23*, 1833, 1841 (1955).
107) Ham, N. S., Ruedenberg, K.: J. Chem. Phys. *29*, 1215 (1958).
108) Dasent, E. W.: Nonexistent compounds. New York: Dekker 1965.
109) Bastiansen, O., Hedberg, L., Hedberg, K.: J. Chem. Phys. *27*, 1311 (1957).
110) Mislow, K., Perlmutter, H. D.: J. Am. Chem. Soc. *84*, 3591 (1962).
111) Watts, L., Fitzpatrick, J. D., Pettit, R.: J. Am. Chem. Soc. *87*, 3253 (1965).
112) Dewar, M. J. S., Kohn, M. C., Trinajstić, N.: J. Am. Chem. Soc. *93*, 3437 (1971).
113) Skell, P. S., Peterson, R. Y.: J. Am. Chem. Soc. *86*, 2530 (1964).
114) Cvetković, D., Gutman, I., Trinajstić, N.: Croat. Chem. Acta *44*, 365 (1972).
115) Longuet–Higgins, H. C.: J. Chem. Phys. *18*, 265 (1950).
116) Jahn, G. A., Teller, E.: Proc. Roy. Soc. (London) *A 161*, 220 (1937).
117) Dewar, M. J. S., Gleicher, G. J.: J. Am. Chem. Soc. *87*, 3255 (1965).
118) Dewar, M. J. S., Harget, A. J., Haselbach, E.: J. Am. Chem. Soc. *91*, 1521 (1969).
119) Wipff, G., Wahlgren, U., Kochanski, E., Lehn, J. M.: Chem. Phys. Letters *11*, 350 (1970).
120) Cvetković, D., Gutman, I.: Matematički Vesnik (Beograd) *9*, 141 (1972).
121) Müller, E., Müller–Rodloff, I.: Liebigs Ann. Chem. *517*, 134 (1935).
122) Dewar, M. J. S., Longuet–Higgins, H. C.: Proc. Roy. Soc. (London) *A 214*, 482 (1952).
123) Cava, M. P., Mitchell, M. J.: Cyclobutadiene and related compounds. New York: Academic Press 1967.
124) Cvetković, D.: private communication, November 1972.
125) Wilcox, C. F.: Tetrahedron Letters *1968*, 795.
126) Bochvar, D. A., Stankevich, I. V.: Zh. Strukt. Khim. *10*, 680 (1969); *12*, 142 (1971); *13*, 1223 (1972).
127) Balaban, A. T.: Paper presented at the 6th Symposium on Theoretical Chemistry, Bad Ischl (Austria) 6—10 April 1970.
128) Živković, T.: Croat. Chem. Acta *44*, 351 (1972).
129) Herndon, W. C.: Tetrahedron *28*, 3675 (1972).
130) Rouvray, D. H.: Compt. Rend. *275 C*, 363 (1972).
131) Fukui, K.: Topics Current Chem. *15/1*, 1 (1970) and references therein. See also Fujimoto, H., Fukui, K.: In: Advances in quantum chemistry, (ed. P.-O. Löwdin) ,Vol. VI, pp. 177. New York: Academic Press 1972.
132) Dewar, M. J. S., Harget, A. J., Trinajstić, N., Worley, S. D.: Tetrahedron *26*, 4505 (1970).
133) Klasinc, L., Trinajstić, N.: Tetrahedron *27*, 4045 (1971).
134) Trinajstić, N.: Record Chem. Progr. *32*, 85 (1971).
135) Young, R. H., Feriosi, D. T.: J. Chem. Soc. Chem. Commun. *1972*, 841.
136) Kearns, D. R.: J. Am. Chem. Soc. *91*, 6554 (1969).
137) Dewar, M. J. S.: J. Chem. Soc. *1952*, 3532.
138) *e.g.* Karrer, P., Jucker, E.: Carotinoide. Basel: Birkhauser 1948.
139) Clar, E.: Polycyclic hydrocarbons. New York: Academic Press 1964.
140) Berlin, A. A., Vinogradov, G. A., Ovchinnikov, A. A.: Intern. J. Quant. Chem. *6*, 263 (1972).

141) For reviews see Sondheimer, F.: Pure Appl. Chem. *7*, 363 (1963); Proc. Roy. Soc. (London) *A 297*, 173 (1967); Proc. Robert A. Welch Found. Conf. Chem. Res. *12*, 125 (1968).
142) Haddon, R. C., Haddon, V. R., Jackman, L. M.: Topics Current Chem. *16*, 103 (1971).
143) Bergman, J.: Nature (London) *194*, 679 (1962).
144) Bergman, J., Hirschfeld, F. L., Rabinovich, D., Schmidt, G. M. J.: Acta Cryst. *19*, 227 (1965).
145) Johnson, S. M., Paul, I. C.: J. Am. Chem. Soc. *90*, 6555 (1968).
146) Milun, M., Sobotka, Ž., Trinajstić, N.: J. Org. Chem. *37*, 139 (1972).
147) Longuet–Higgins, H. C., Salem, L.: Proc. Roy. Soc. (London) *A 251*, 172 (1959); *A 257*, 445 (1960).
148) Sondheimer, F., Wolovsky, R.: J. Am. Chem. Soc. *84*, 260 (1962).
149) Sondheimer, F., Gaoni, Y.: J. Am. Chem. Soc. *83*, 1259 (1962).
150) Dewar, M. J. S., Gleicher, G. J.: J. Am. Chem. Soc. *87*, 685 (1965).
151) Fukui, K., Imamura, A., Yonezawa, T., Nagata, C.: Bull. Chem. Soc. Japan *33*, 1591 (1960).
152) Fukui, K., Fujimoto, H.: Bull. Chem. Soc. Japan *40*, 2024 (1967).
153) Gutman, I., Milun, M., Trinajstić, N.: J. Chem. Phys., in press.
154) Goldstein, M. J., Hoffmann, R.: J. Am. Chem. Soc. *93*, 6193 (1971).
155) Woodward, R. B., Hoffmann, R.: The conservation of orbital symmetry. Weinheim: Verlag Chemie GmbH 1970.
156) van der Hart, W. J., Mulder, J. J. C., Oosterhoff, L. J.: J. Am. Chem. Soc. *94*, 5724 (1972).
157) Salem, L.: Chem. Brit. *5*, 449 (1969).
158) Trindle, C.: J. Am. Chem. Soc. *92*, 3251 (1970); Theoret. Chim. Acta *18*, 261 (1970).
159) See for review Dewar, M. J. S.: Angew. Chem. intern. Ed. Engl. *10*, 761 (1971).
160) Ugi, I., Marquarding, D., Klusacek, H., Gokel, G., Gillespie, P.: Angew. Chem. *82*, 741 (1970).
161) Doering, W. von E., Knox, L. H.: J. Am. Chem. Soc. *76*, 3203 (1954). — Dewar, M. J. S., Pettit, R.: Chem. Ind. (London) *1955*, 199; J. Chem. Soc. 2021, *1956*, 2026. — Dauben, H. J., Gadecky, F. A., Harmon, K. M., Pearson, D. L.: J. Am. Chem. Soc. *79*, 4557 (1957). — Pauson, P. L.: In: Non-benzenoid aromatic hydrocarbons (ed. Ginsburg, D.), p. 107. New York: Interscience Publ. 1959.
162) Sondheimer, F., Calder, I. C., Elix, J. A., Gaoni, Y., Garratt, P. J., Grohman, K., di Maio, G., Mayer, J., Sargent, M. V., Wolovsky, R.: Chem. Soc. (London), Spec. Publ. *21*, 75 (1967).
163) Bochvar, D. A., Tutkevich, A. V.: Izv. Akad. Nauk USSR, Ser. Khim. *1966*, 756.
164) Bochvar, D. A., Stankevich, I. V., Tutkevich, A. V.: Izv. Akad. Nauk USSR, Ser. Khim. *1969*, 1185.
165) Badger, G. M.: Aromatic character and aromaticity, Chap. 4. Cambridge: University Press 1969.
166) Some work is at present being carried out in our laboratory.
167) Gutman, I., Trinajstić, N.: Chem. Phys. Letters *17*, 535 (1972), *20*, 257 (1973).
168) Stepanov, N. F., Tatevskii, V. M.: Zh. Strukt. Khim. *2*, 204, 452 (1961).
169) Tatevskii, V. M.: Zh. Fiz. Khim. *34*, 241 (1960).
170) McClelland, B. J.: J. Chem. Phys. *54*, 640 (1971).
171) Brown, R. D.: Trans. Faraday Soc. *46*, 1013 (1950).
172) Sahini, V. E.: J. Chim. Phys. *59*, 177 (1962); Rev. Chim., Acad. Rep. Populaire Roumaine *7*, 1265 (1962); Rev. Chim. (Bucharest) *15*, 551 (1964).

173) Hakala, R. W.: Intern. J. Quant. Chem. *1S*, 187 (1967).
174) Balaban, A. T.: Rev. Roumaine Chim. *15*, 1243 (1970).
175) Smith, W. B.: J. Chem. Educ. *48*, 749 (1971).
176) Baird, N. C.: Can. J. Chem. *47*, 3535 (1969); J. Chem. Educ. *48*, 509 (1971).
177) Tatevskii, V. M.: Zh. Fiz. Khim. *25*, 211 (1951).
178) Green, A. L.: J. Chem. Soc. *1956*, 1886.
179) Wheland, G. W.: The theory of resonance and its application to organic chemistry. New York: Wiley 1953.
180) Pauling, L.: J. Chem. Phys. *1*, 280 (1933).
181) Pauling, L., Wheland, G. W.: J. Chem. Phys. *1*, 362 (1933).
182) Pauling, L., Sherman, J.: J. Chem. Phys. *1*, 679 (1933).
183) Rumer, G.: Göttingen. Nachr. *337* (1932).
184) Longuet–Higgins, H. C.: J. Chem. Phys. *18*, 275, 283 (1950).
185) Heilbronner, E.: Helv. Chim. Acta *45*, 1722 (1962).
186) Platt, J. R.: In: Encyclopedia of physics (ed. S. Flügge), Vol. 37, pp. 173. Berlin–Heidelberg–New York: Springer 1961.
187) Cvetković, D., Gutman, I., Trinajstić, N.: To be published.
188) Marcus, M., Minc, H.: Am. Math. Monthly *72*, 577 (1965).
189) Schmidtke, H. H.: Coord. Chem. Rev. *2*, 3 (1967).
190) Schmidtke, H. H.: Intern. J. Quant. Chem. *2S*, 101 (1968).
191) Schmidtke, H. H.: J. Chem. Phys. *48*, 970 (1968).
192) Kettle, S. F. A.: Theoret. Chim. Acta *3*, 211 (1965); *4*, 150 (1966).
193) Kettle, S. F. A., Tomlinson, V.: J. Chem. Soc. *A 1969*, 2002, 2007; Theoret. Chim. Acta *14*, 175 (1969).
194) Rudolph, R. W., Pretzer, W. R.: Inorg. Chem. *11*, 1974 (1972).
195) Gutman, I., Trinajstić, N.: to be published.
196) Stankevich, I. V.: Zh. Fiz. Khim. *42*, 1876 (1968).
197) Stankevich, I. V.: Zh. Fiz. Khim. *43*, 549, 556 (1969).
198) Stankevich, I. V.: Zh. Fiz. Khim. *44*, 1540 (1970).
199) Stankevich, I. V.: Zh. Fiz. Khim. *46*, 2463 (1972).
200) Graovac, A., Gutman, I., Trinajstić, N.: In preparation.
201) See also Brown, R. D.: J. Chem. Soc. *1953*, 2615.

Received January 5, 1973

The Electrostatic Molecular Potential as a Tool
for the Interpretation of Molecular Properties

Prof. Dr. Eolo Scrocco and **Prof. Jacopo Tomasi**

Laboratorio di Chimica Quantistica ed Energetica Molecolare del CNR, Pisa, Italy

Contents

I.	Introduction	97
II.	Limits of Electrostatic Approximation in Molecular Interaction Problems	98
	A. Preliminary Remarks	98
	B. Born-Oppenheimer Approximation	99
	C. SCF Approximation	99
	D. Hartree Approximation	101
	E. Electrostatic Approximation	102
III.	The Electrostatic Molecular Potential	104
IV.	The Variety of Shapes of the Electrostatic Potential	106
	A. Introduction	106
	B. Saturated Compounds	107
	1. Water and Ammonia	107
	2. Cyclopropane and Derivatives	109
	C. Unsaturated Compounds	116
	1. Cyclopropene and Derivatives	116
	2. Nitrogen Molecule	119
	3. Triatomic Molecules: O_3 and FNO	120
	4. Formamide	121
	D. Heteroaromatic Compounds	123
	1. Five-membered Rings	123
	2. Six-membered Heterocycles	130
	3. Purinic Bases	133
	E. Provisional Conclusions	135

E. Scrocco and J. Tomasi

V.   The Dependence of $V(r)$ on the Accuracy of the Wave Function 135

       A. SCF Wave Functions 135

       B. Semi-empirical Wave Functions 138

VI.   Protonation Processes 139

VII.   Group Contributions to the Electrostatic Potential 143

       A. Localized Orbitals and Related Partition of $V(r)$ 143

       B. An Example of Analysis of $W(r)$ 145

       C. Conservation Degree of Group Potentials 149
          1. CH_2 Group 150
          2. NH and C-C Groups 153

VIII. Analytical Approximations of $V(r)$ 153

       A. One- and Many-center Multipole Expansions 153

       B. Monopole Expansions 156

IX.   Electrostatic Description of the Conformational Structures of Molecular
       Associates $A \cdot H_2O$ 157

       A. Direct Application of the SCF Electrostatic Potential 157

       B. Application of the Analytical Expansions of $V(r)$ 162

X.   References 167

I. Introduction

The increasing efficiency of large computers permits more and more extensive utilization of the methods of quantum chemistry to shed light on both static and dynamic properties of small and medium-sized molecules.

However, the computational effort increases rapidly with the number of atoms involved and reaches such prohibitive levels that the chemist's desire to have, on theoretical grounds, a reasonably accurate forecast of molecular interactions and chemical reactivity has so far proved unrealizable. It is hence necessary to resort to approximate methods which yield at least qualitative indications and a rough prediction of the phenomenology involved.

Examples of such approximate methods are the computation of intermolecular interaction energies in terms of experimental multipole moments of empirical atomic contributions.

Another category is represented by some current reactivity theories which rely upon a large set of molecular indices, such as atomic populations, bond orders, free valency, autopolarizability, etc. These data represent an attempt to extract from the properties of the isolated molecule some useful information about its behavior as it interacts with other molecules.

In the same spirit, we report here an attempt to utilize for the study of molecular interactions the analysis of the electrostatic potential (produced in the surrounding space) which can be calculated from the wave function of the isolated molecule. The electrostatic molecular potential is generally a rather complex function of the point, and for this reason much of the material is presented in graphic form, as this permits a quick and easy visualization of the outstanding features, although some emphasis is also given to analytic representations of the electrostatic potential as well as to their convergence properties.

An attempt is made to utilize the electrostatic potential for a first-order prediction of the relative reactivity of functional groups in ionic reactions and to characterize such groups according to the shape of the potential in the corresponding portion of the outer molecular space. The molecular potential can then be broken down into contributions due to the different groups present in the molecule and the resulting analysis will give an idea of the degree of *conservation* and *transferability* of group electrostatic potential among chemically related molecules.

Lastly, we describe a method which utilizes the electrostatic potential for a first-approximation study of the energetics of conformational interactions between organic molecules (especially those containing heteroatoms) and small polar molecules like water. This is clearly relevant to solvation, but no attempt is here made to extrapolate the method to larger assemblies of molecules.

II. Limits of Electrostatic Approximation in Molecular Interaction Problems

A. Preliminary Remarks

This section illustrates the significance and limits of electrostatic approximation by listing a series of successive decreasing-order approximations which, on past experience, may be considered to deal reasonably with a typical problem of chemical interaction.

The researcher is mainly interested in investigating the mechanism of the overall reaction. He must call upon his experience and ingenuity in order to:

i) elaborate a model which replaces the real system without any loss of major characteristics;

ii) select the appropriate level of approximation and adjust the program to the actual case.

The first point is especially delicate. The choice of a suitable model requires that the reaction mechanism be stated with precision; once the model has been chosen, the whole problem is reduced to studying one at a time certain reactions which in the overall process either succeed or compete with each other to give a range of possible products. The quality of the results obtained is an "a posteriori" test of our insight of physical reality.

We shall not dwell on this topic; we limit our considerations to a very simple model for a given reaction, namely the interaction between two single molecules A and B.

To study an elementary reaction rigorously, we would require a compete knowledge of the time evolution of the system in question. According to the laws of quantum mechanics, such an approach would require the determination of the overall wave function, explicit in all the coordinates. Clearly, it is not easy to fulfil such exacting prescriptions, and indeed they can only be satisfied for very simple cases.

For the purposes of this paper, we shall consider only cases where A and B are both closed shell systems, and where one of them (the reactant) is charged, or is a small molecule having a noticeable dipole moment. Attention will be focused on the level of approximation appropriate to deal with medium- and long-range interactions, and we will be satisfied if we succeed in detecting the best approach channels for the reactant.

B. Born-Oppenheimer Approximation

A particularly convenient approximation consists in separating the nuclear motions from the overall Schrödinger equation (Born-Oppenheimer approximation). The problem is then reduced to a search for the stationary electronic states with energy $E(\boldsymbol{R})$ for fixed values of the set of nuclear coordinates $\boldsymbol{R}$ and the calculation of the corresponding electrostatic nuclear repulsion energy $V_n(\boldsymbol{R})$. Varying $\boldsymbol{R}$ gives an energy hypersurface, $W(\boldsymbol{R}) = E(\boldsymbol{R}) + V_n(\boldsymbol{R})$, which has as many dimensions as there are parameters necessary to specify the nuclear geometry. $W(\boldsymbol{R})$ is then inserted in the nuclear Hamiltonian as a potential energy term; if the coupling between nuclear and electronic motions is neglected, the problem of determining the time evolution of the system is reduced to the study of the motion of a representative point on the $W(\boldsymbol{R})$ surface.

C. SCF Approximation

The introduction of the Born-Oppenheimer approximation is not sufficient to make the problem actually solvable. To determine the electronic wave function — a necessary step to construct the potential hypersurface $W(\boldsymbol{R})$ — we have to resort to further approximations

For the case of interactions between closed shell systems (like those considered in the present paper), a sufficient approximation is offered by the one-determinant SCF wavefunction $\Psi_{AB}^{SCF}(\boldsymbol{r},\boldsymbol{R})$[a]. We suppose that all our readers are sufficiently well acquainted with SCF theory and we will not repeat here an exposition of the basic procedure, which may be found in all textbooks on quantum chemistry. We simply mention a few points which will be useful later on. The SFC wave function for the $2N$ electrons of the AB system:

$$\Psi_{AB}^{SCF} = [1/(2N)!]^{\frac{1}{2}} \det |\varphi_1(\boldsymbol{r}_1)\alpha_1 \varphi_1(\boldsymbol{r}_2)\beta_2 \varphi_2(\boldsymbol{r}_3)\alpha_3 \ldots \ldots \varphi_N(\boldsymbol{r}_{2N})\beta_{2N}|$$

$$(1)$$

(φ_l is a molecular orbital depending only on the spatial coordinates $\boldsymbol{r}_l \equiv x_l$, y_l, z_l of the l-th electron, and α and β are spin eigenfunctions) is usually

---

[a] The errors introduced by neglecting the electronic correlation, which are inherent to the one-determinant approximation, will be largely overcome by the uncertainties arising from other subsequent approximations, so that it is of little use to introduce a correction for this type of error.

obtained by considering the φ_i's as expressed in the form of linear combinations

$$\varphi_i = \sum_k c_{ki}\, \chi_k \tag{2}$$

of a set of n basis functions

$$\chi = (\chi_1,\, \chi_2 \ldots \cdot \chi_n) \tag{3}$$

centered at the nuclei of the AB system (MOLCAO approximation). The c coefficients of the expansion (2) are determined by an iterative solution of the Hartree-Fock equations which requires a sequence of diagonalizations of matrices of order n.

Once Ψ_{AB}^{SCF} is calculated (for the given configuration of the nuclei), the total energy E is obtained as a sum of nucleus—nucleus repulsions, nucleus—electron attractions (not critical from the point of view of the present discussion, since they are easy to compute) and two-electron repulsion contributions which can again be divided into coulombic, J, and exchange terms, K:

$$J = \sum_{i,j} (\varphi_i\varphi_i|\varphi_j\varphi_j) = \sum_{i,j} \sum_{r,s} \sum_{t,v} c_{ri}^* c_{si} c_{tj}^* c_{vj} (\chi_r \chi_s | \chi_t \chi_v) \tag{4}$$

$$K = \sum_{i,j} (\varphi_i\varphi_j|\varphi_i\varphi_j) = \sum_{i,j} \sum_{r,s} \sum_{t,v} c_{ri}^* c_{si}^* c_{tj} c_{vj} (\chi_r \chi_t | \chi_s \chi_v) \tag{5}$$

The electron repulsion integrals have also been expressed in terms of the expansion basis functions χ to show that the calculation of the energy requires previous knowledge of all the elements $I_{rs,tv} = (\chi_r \chi_s | \chi_t \chi_v)$ of a supermatrix I of $n^2 \times n^2$ dimension [b]. Such a scheme of calculation, including the I matrix, must be repeated for each point on the nuclear conformation hypersurface.

Finally, the calculation of a potential energy surface $W(R)$ requires a considerable computational effort which rapidly becomes prohibitive if the number of atoms and electrons included in the model exceeds a very low threshold. The majority of problems of chemical interest cannot be yet treated in this way.

---

[b] For symmetry reasons, no more than $n^4/8$ integrals have actually to be computed. The manipulation and diagonalization of the Hartree-Fock matrices increases the computational effort by a factor of n^5. Thus, the related overall rate can be assumed to be proportional to $n^{4,5}$, as suggested by Boys and Rajagopal[1].

D. Hartree Approximation

Since a way must be found to overcome the above-mentioned technical difficulties, let us focus our attention on the more specific aspects of the problem and to seek a solution for that particular subject.

Let us first take hypersurface regions corresponding to relatively large distances between A and B, where the two molecules retain their internal structure. Further, we will try to maintain the hypothesis of internal nuclear rigidity for a limited range of shorter distances. This restriction, of course, excludes from our treatment the part of the reaction

$$A + B \rightarrow AB \rightarrow products$$

which leads to the rearrangement of the atoms of A and B and, ultimately, to the reaction products. In spite of these limitations, the investigation of the initial parts of the reaction channels is worthwhile; moreover, there is a reasonable hope that, for some families of reactions, such a study can also give some useful information on the intermediate complex AB.

It is well known that the exchange contributions to the energy decrease with distance more rapidly than the Coulomb ones. Thus, it appears sound to employ, for large portions of the energy hypersurface, an approximate treatment which preserves the features of the SCF method while using approximate expressions for the exchange terms $K_{iA,jB} = (\varphi_{iA}\varphi_{jB}|\varphi_{iA}\varphi_{jB})$ between the molecular orbitals of A and B. As a limit approximation, such exchange terms could be completely neglected in the calculations. A direct introduction of this last approximation in the SCF MOLCAO framework is not particularly fruitful because it does not change the number of the I supermatrix elements to be calculated and does not reduce the dimension of the Hartree-Fock matrix. It does, however, open the way to another, more remunerative simplification, which takes advantage of the fact that, in the portions of hypersurface where exchange terms are negligible, one can safely neglect the charge transfer between molecules. In this case the two fragments A and B of the whole system have a more evident individuality: the number of electrons is clearly defined in both partners, which can be rightly considered as individual molecules. The requirement of expanding the molecular orbitals φ_{iA} pertaining to molecule A on the overall set $\chi = \chi_A + \chi_B$ is no longer necessary and the expansion can be reduced to the subset χ_A of extension $n_A < n$. (Analogous remarks apply, of course, to the φ_B orbitals of B.) In other words, this approximation leads to a mere factorization of the electronic wave function of the system:

$$\Psi_{AB}^H = \Psi_A^H \cdot \Psi_B^H \tag{6}$$

101

where the two one-determinant wave functions Ψ_A^H and Ψ_B^H are now expanded in the two subsets χ_A and χ_B, respectively, which are approximately to be considered as orthogonal. The number of two-electron integrals to be calculated is thus much reduced: one needs only the portions of the I supermatrix corresponding to elements of the following types:

$$(\chi_{rA}\,\chi_{sA}\,|\,\chi_{tA}\,\chi_{vA}),\ (\chi_{rB}\,\chi_{sB}\,|\,\chi_{tB}\,\chi_{vB}),\ (\chi_{rA}\,\chi_{sA}\,|\,\chi_{tB}\,\chi_{vB})$$

without any other arrangement of the subindexes A and B. Moreover, the order of the Hartree-Fock secular equation is also reduced since it is factorized in two blocks whose dimensions are respectively n_A and n_B, $(n_A + n_B = n)$.

Eq. (6) may be considered as a simplified representation of the AB system in the context of the group function method[2]: $\Psi_{AB} = \mathscr{A}\Psi_A\Psi_B$. Wave function (6) partially violates the Pauli exclusion principle because the antisymmetrizer $\mathscr{A}$ acting on the product between Ψ_A and Ψ_B (singly antisymmetric) is missing. This approximation is parallel to the one Hartree introduced for atomic calculations which is why it is called the molecular Hartree approximation.

The variational iterative procedure to optimize wave function (6) is performed by an alternating process, which consists in first bringing to selfconsistency Ψ_A in the electrostatic field of molecule B (in addition, of course, to its own field of electrons and nuclei), then Ψ_B in the field of A, and so on until convergence is reached. The energy thus obtained contains the Coulomb interaction and polarization terms between A and B — *i.e.* the most important terms at large separation — and discards charge-transfer and dispersion effects.

The nomenclature used in this section concerning the partition of the interaction energy into separate effects is borrowed from an alternative perturbation approach to the problem we are discussing[3]. Perturbative treatments are, in fact, particularly effective in the portions of the hypersurface we are presently considering. The next step in our approximation scheme may be considered also a first-order perturbative treatment.

E. Electrostatic Approximation

In the hypersurface portions where polarization effects may be considered inessential to the understanding of the physical phenomenon under investigation, the model may be limited to electrostatic interactions. This further reduction is less well justified than the preceding ones, although in some cases (see, *e.g.* Section IX) a mutual cancellation of other effects enhances the reliability of the electrostatic approximation.

In any case, before assuming its validity, this approximation must be carefully controlled.

Purely electrostatic interactions are taken into account by another even more simplified wave function, expressed as the simple product

$$\Psi^O_{AB} = \Psi^O_A \cdot \Psi^O_B \tag{7}$$

of two antisymmetrized wave functions, Ψ^O_A and Ψ^O_B, which are simply the SCF wave functions of the isolated molecules A and B. According to this approximation, the interaction energy is given by:

$$W_{AB} = E^O(AB) - [E^O(A) + E^O(B)] =$$

$$= -2\sum_i^B \sum_\beta (\varphi_{iA}(1)\varphi_{iA}(1) \left| \frac{Z_a}{r_{1\beta}} \right.) - 2\sum_j^A \sum_a (\varphi_{jB}(2)\varphi_{jB}(2) \left| \frac{Z_a}{r_{2\beta}} \right.) +$$

$$+ 4\sum_i \sum_j (\varphi_{iA}(1)\varphi_{iA}(1) | \varphi_{jB}(2)\varphi_{jB}(2)) + \sum_a^A \sum_\beta^B \frac{Z_a Z_\beta}{R_{a\beta}} \tag{8}$$

The first two terms respectively give the attraction energy among the electrons of A and the nuclei of B (having charges equal to Z_β), and among the electrons of B and nuclei of A. Both terms are expressed as a sum of one-electron integrals. The third term of (8) provides the repulsion among electrons of A and B (two-electron integrals) while the last term gives the repulsion among the nuclei of the two molecules. Only a portion of the I supermatrix is needed to calculate W_{AB} according to Eq. (8), *i.e.* that part corresponding to Coulomb integrals $(\chi_{rA} \chi_{sA} | \chi_{tB} \chi_{vB})$ among basis functions pertaining to the subsets χ_A and χ_B. In addition, diagonalizations of Hartree-Fock matrices are no longer necessary.

Since we have now attained the level of approximation which is the subject of the present paper, we shall consider the topic in more detail.

One may regard W_{AB}, Eq. (8), as the interaction energy between the potential field $V_A(\boldsymbol{r})$ arising from the first charge distribution $\gamma_A(\boldsymbol{r}_1)$, and the second charge distribution γ_B. It could seem idle to speculate which of the two partners will be described by the charge distribution and which by the potential field. However, in reactivity problems one is mainly concerned with relatively large molecules interacting with simpler reactants, so we will adopt, as most convenient, the convention of calculating W_{AB} in terms of the electrostatic potential of the relevant molecule (V_A) and of the charge distribution of the reactant (γ_B). It is useful to state this convention at this point because it is needed for a further approximation step we introduce in order to cut out the remaining two-electron integrals of Eq. (8).

It is convenient to keep V_A as accurate as possible so that it can be calculated directly from Ψ_A^0 without other approximations (see, however, Section VIII for analytical expansions of V_A) and to limit further simplifications to γ_B.

In the range of applications of the electrostatic method we have attempted, it was found useful to approximate γ_B by a set of suitably placed point charges q_{kB} (this approximation has given quite good results both for small neutral molecules having a noticeable dipole moment, like H_2O and NH_3, and for molecular ions like NO_2^-). In the case of atomic ions, the point charge set may be reduced to one charge only. Within this approximation, the calculation of W_{AB} is much simplified:

$$W_{AB} = \sum_k V_A(k) q_{kB}(k) \tag{9}$$

The two-electron integrals are no longer necessary and a good share of the information relative to the chemical process considered is contained in the V_A function which deserves to be studied and analyzed *"per se"*.

III. The Electrostatic Molecular Potential

The electrostatic potential arising form molecule A is completely defined at every point of the space if one knows the charge distribution (electronic and nuclear) of the molecule:

$$\gamma_A(r_1, R) = -\varrho_A(r_1) + \sum_a Z_a \delta(r_1 - R_a) \tag{10}$$

In Eq. (10) $\varrho_A(r_1)$ represents the electron charge distribution, *i.e.* the diagonal element of the first-order electron density matrix which, in the SCF approximation [c], is given by

$$\varrho_A(r_1) = 2 \sum_i \varphi_{iA}^*(r_1) \varphi_{iA}(r_1) \tag{11}$$

this, in turn, when expanded in terms of an atomic basis (MOLCAO approximation), becomes

$$\varrho_A(r_1) = \sum_r \sum_s P_{rs} \chi_r^*(r_1) \chi_s(r_1) \tag{12}$$

---

[c] It is of course, possible, to go beyond the SCF approximation and to use more general expressions of ϱ_A.

where $P_{rs} = 2\sum_{i} c_{ri}c_{si}$ is the rs-th element of the population matrix. The discrete point charge distribution of the nuclei is symbolically represented in Eq. (10) by a sum over continuous Dirac delta functions written in terms of the same running variable r_1 as for electrons.

The electrostatic potential at point r is given by

$$V(r) = \int \frac{\gamma_A(r_1)}{|r - r_1|}\, dr_1 = \int \frac{\varrho_A(r_1)}{|r - r_1|}\, dr_1 + \sum_{a} \frac{Z_a}{|r - R_a|}$$

$$= -\sum_{r}\sum_{s} P_{rs} \int \frac{\chi_r^*(r_1)\, \chi_s(r_1)}{|r - r_1|}\, dr_1 + \sum_{a} \frac{Z_a}{|r - R_a|} \tag{13}$$

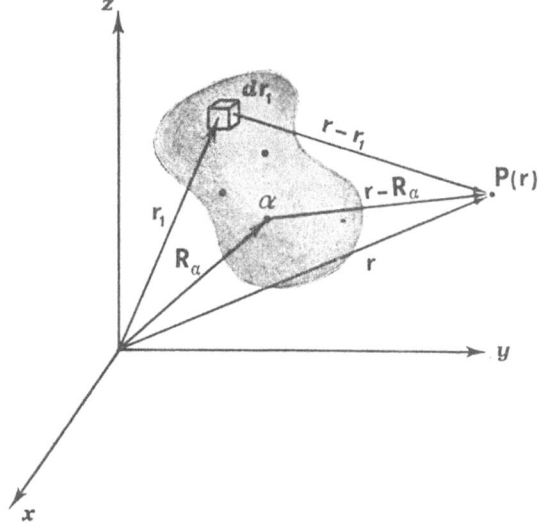

Fig. 1. Notations adopted in the definition of the electrostatic potential in a point $P(r)$ generated by the electronic and nuclear charges of a molecule

and its calculation requires only the evaluation of one-electron integrals over the selected expansion basis. A large number of computer routines is at present available for such integrals, but in some programs obtaining $V(r)$ is even more straightforward because it is a byproduct of the calculation of energy, $V(r)$ being related to the first integration step of two-electron repulsion terms[4]. As stated above, the definition (13) of the electrostatic potential does not require ϱ_A to be expressed in the SCF framework, but it is convenient to point out that $V_A(r)$ is the expectation value of a one-electron operator so that, according to the Brillouin theorem, its SCF approximation is correct to one order higher than the SCF wave function employed.

$V(r)$ also represents the value, at the first order of perturbation, of the interaction energy of molecule A with a unitary point charge (*e.g.* a proton).

The electrostatic molecular potential may be considered, therefore, from two points of view: 1. as an expectation value, which is accordingly clearly defined whatever method was employed to calculate the wave function (*e.g.* one-determinant SCF, many-determinant C.I., etc.) and whatever approximation level was maintained in the computation (*e g.* kind and extension of the expansion basis χ, etc.) and 2. as an approximation, at a clearly defined order, of the interaction energy within a system of point charges (Eq. 9).

Both aspects will be taken into account in our analysis: these should provide, on the one hand, a visualization of the features of molecular charge distribution — *i.e.* comparisons and relationships among different molecules or among similar chemical groups placed in different chemical frames — and, on the other hand, an approximate picture of the capability of the molecule in question to interact with other chemical species. The more correct the first-order approximation, the sharper this picture becomes. It is particularly well suited for regions at medium or large distances from the molecule where reaction channels begin to assume a definite shape.

IV. The Variety of Shapes of the Electrostatic Potential

A. Introduction

Some chemically relevant examples of characteristic shapes of the molecular electrostatic potential are described here and presented graphically by means of isopotential curves drawn in selected planes. More precisely, the quantity reported in the maps is not the potential $V(r)$, but the interaction energy $W(r)$ of this potential with a positive unit charge ($+e$). This presentation is better because it gives directly the interaction energy (at the first order) of the molecule in question with electrophilic agents. Since chemical interactions are involved, the "chemical" unit of energy, kcal/mole, is used instead of the atomic units employed in the preceding section. For clarity, the term "electrostatic potential" will be retained, the conversion to energy being a mere fact of presentation.

The examples will be presented in the following order:

i) Saturated compounds

    a) H_2O and NH_3 as examples of monofunctional molecules

    b) cyclopropane and allicyclic derivatives as examples of poly-functional molecules.

ii) Unsaturated compounds: miscellaneous examples containing the groups C=C, C=O, $-N=O$, $-N=N-$, etc.

iii) Heteroaromatic compounds

    a) Five-membered cycles

    b) Six-membered cycles

    c) Bases of the nucleic acids

A roman reference number is appended to each molecule. Not all the electrostatic potential calculations at present available are discussed here, though other examples are considered in the next section. The calculations of $W(r)$ for molecules XII, XV, XX, XXI are presented for the first time. The others have been published before, but some of the maps have been drawn specially for this paper. The kind permission of authors and editors to reproduce some material is here acknowledged.

B. Saturated Compounds

1. Water and Ammonia

Potential energy maps for water (I) are reported in Figs. 2 and 3. Fig. 2 refers to the molecular plane and Fig. 3 to the second symmetry plane perpendicular to the molecule. Potential values are higher and positive in the proximity of atoms, where the nuclear charges are only partially shielded by the electron cloud. Such portions of space, which we could call inner molecular space [d], are not of primary interest in the present review.

The outer molecular space is partitioned by nodal surfaces into different portions. In the molecular plane on the side of the hydrogens, there is a positive region where the approach of a positively charged reactant is disfavored. On the opposite side $W(r)$ is negative and the approach of a positive charged reactant is favored. The anisotropy of the potential is similar to that produced by a dipole, but the picture given

---

[d] The division into inner and outer molecular space is highly empirical: as a sort of guide one could consider the demarcation line as being placed somewhere between covalent and van der Waals surfaces.

by $W(\mathbf{r})$ is, in fact, somewhat more complex; on comparing Figs. 2 and 3 one may note how in the negative region there is a large attractive hole which extends above and below the molecular plane and which contains two minima.

The shape of the molecular potential reflects the generally accepted intuitive picture of the charge distribution of the H_2O molecule: four electrons interested in two O—H bonds and two electron lone pairs,

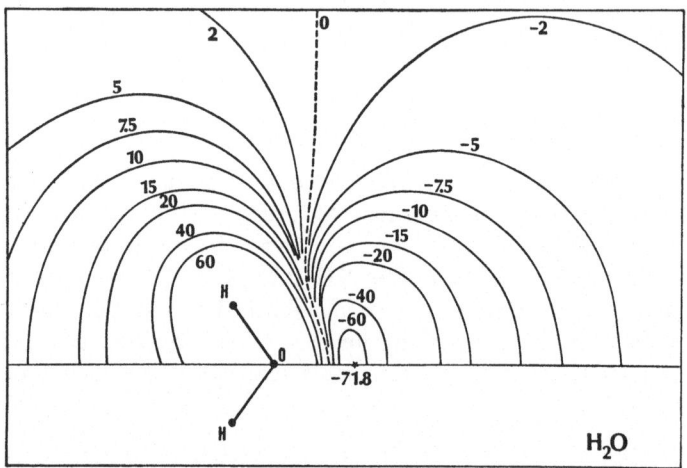

Fig. 2. Electrostatic potential-energy map for H_2O (I) in the molecular plane. Values are expressed in kcal-mole

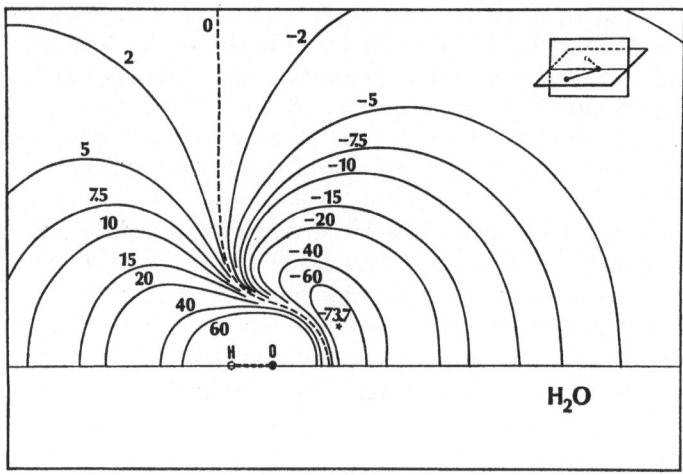

Fig. 3. Potential-energy map for H_2O in the symmetry plane perpendicular to the molecular one

equivalent among them, placed in the symmetry plane orthogonal to the O—H bonds and with nucleophilic character.

For the ammonia molecule (II), isoelectronic with water, the electrostatic potential has a shape which is shown in part in Fig. 4[5]. The figure

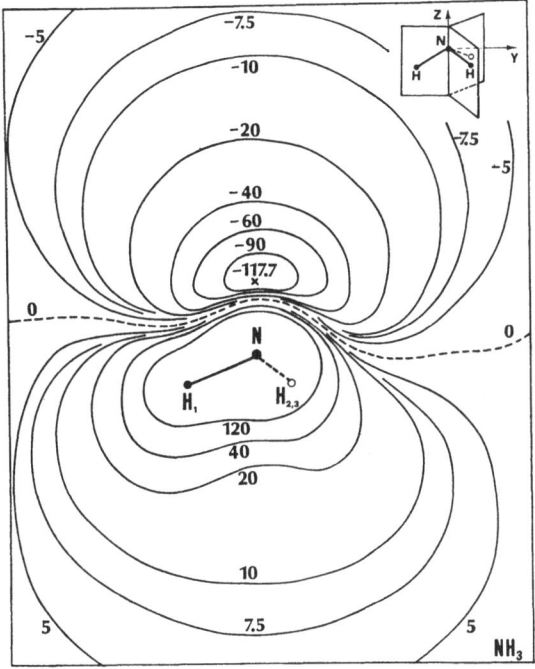

Fig. 4. Potential-energy map for NH₃ (II) in a symmetry plane. From Ref. [5]

refers to one of the three equivalent symmetry planes. In this case too, a positive region correspond to the N—H bonds, and a negative one, less extended than for H_2O and without indication of double minimum, is present. Such behavior is in accordance with classical descriptions which assign to NH_3 only a lone pair.

The values of $W(r)$ in the minima are quite different in H_2O and NH_3: —73.7 kcal/mole as against —117.7 kcal/mole. Rather than taking absolute values, it is better to pay attention to this difference, which is in accordance with the differences in proton affinity of such simple molecules. We will return to this topic later.

2. Cyclopropane and Derivatives

No potential energy maps of simple organic compounds deriving from water and ammonia by single substitutions are at present available.

Our comparative examination of electrostatic potential will be continued on the cyclopropane family.

Cyclopropane and derivatives have a special structural peculiarity: the three-atom ring structure induces a bending in the ring bonds[e]. Fig. 5 reports the $W(\mathbf{r})$ map for the ring plane of cyclopropane (III).

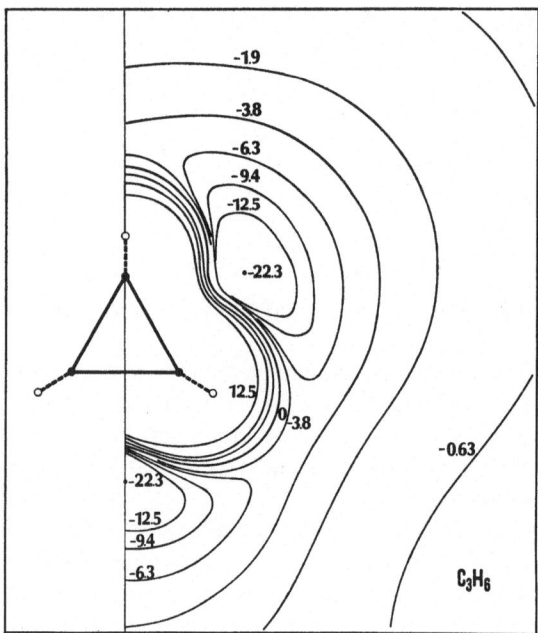

Three minima are seen, symmetrically placed and corresponding to the three C—C bent bonds, and correspondingly in the space surrounding the molecule three channels of approach for electrophilic reactants are

Fig. 5. Potential-energy map for cyclopropane (III) in the ring plane. From Ref. [9]

---

[e] The experimental evidence of the bending of ring bonds[6] can be supported in terms of atomic hybrids[7] and in terms of sigma-pi orbitals[8]. Such interpretations of electronic structure can be verified also in terms of SCF localized orbitals[9].

well in evidence. These channels ultimately lead to the minima and are separated by space portions where the repulsive potential of the CH_2 groups is greater. Such electrostatic description agrees with the 'ab initio' calculations[10] in forecasting as favored, along the protonation process, the intermediate form III a, rather than the alternative face-protonated form. By the way, it is convenient to recall that the electrostatic approximation may give some information on the initial stage of the approach of molecules A and B, and that in some cases this information can be extrapolated to the "supermolecule" AB. Conversely, it is not possible to obtain from the electrostatic approximation elucidations of any sort concerning further steps in the reaction.

As far as cyclopropane is concerned, extensive studies[11] on the relative stability different geometries of $C_3H_7^+$ show that open forms of the cation are the stable ones. Such findings do not contradict the hypothesis that the proton approach channel is really of the type suggested by the electrostatic picture. From the form III a the cation may pass to more stable geometries through further rearrangement channels which seem to have not too large energy barriers[11].

Passing now to cyclopropane derivatives, we examine first some potential maps of oxirane (IV)[9] and aziridine (V)[9].

IV   V   V a

Figs. 6 and 7 refer to the ring plane of oxirane and its second symmetry plane, respectively (molecular symmetry group C_{2v}). Both figures also contain the perpendicular projection of the nuclei placed outside the above planes (blank circles). The substitution of a CH_2 group with an oxygen leads to appreciable variations in the potential. The minimum related to the remaining C—C bond is smaller than in cyclopropane (O is more electronegative than CH_2). The minima of the other two bent bonds disappear, being absorbed by a unique negative region around the heteroatom. Such a region, as may be seen from Fig. 7, is more structured than the corresponding one in H_2O. Here we have two well-evidenced minima separated by a barrier of about 9 kcal/mole. The shape of the potential corresponds for this molecule too, to the classic description of the electronic structure: the location of the two minima is in reasonable agreement with the direction of the two oxygen lone pairs.

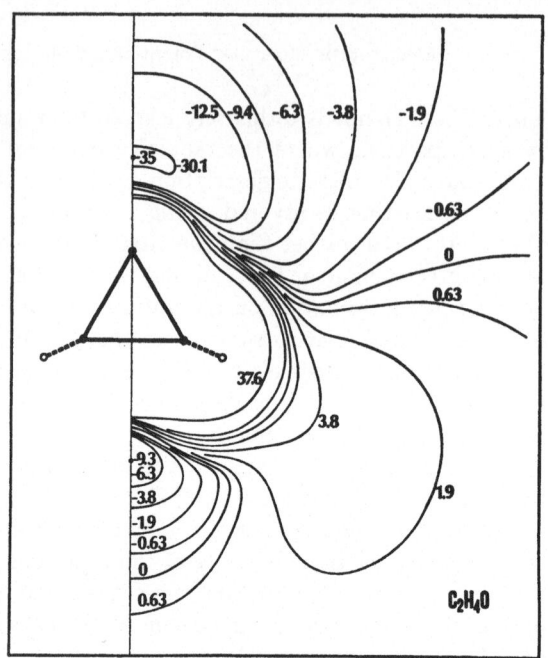

Fig. 6. Potential-energy map for oxirane (IV) in the ring plane. From Ref. [9]

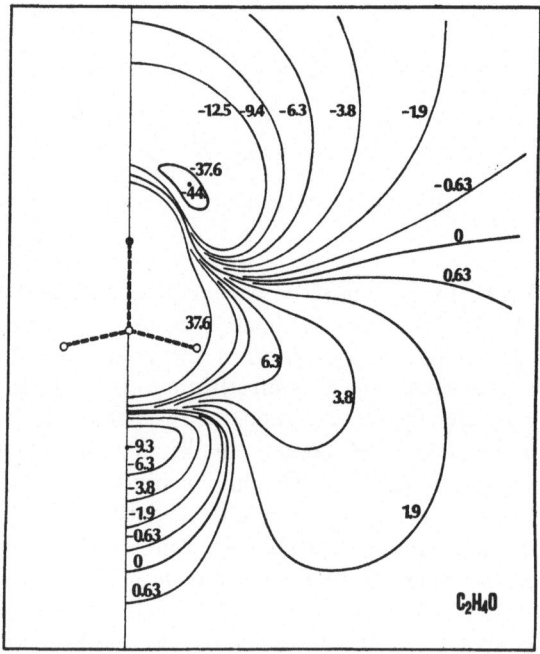

Fig. 7. Potential-energy map for oxirane in the symmetry plane perpendicular to the ring. From Ref. [9]

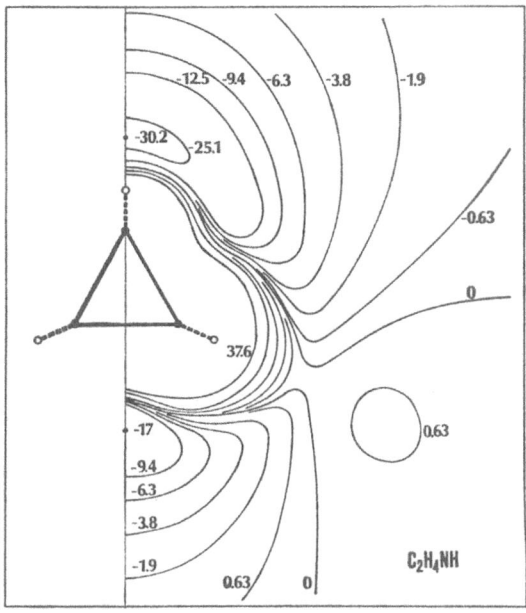

Fig. 8. Potential-energy map for aziridine (V) in the ring plane. From Ref. [9]

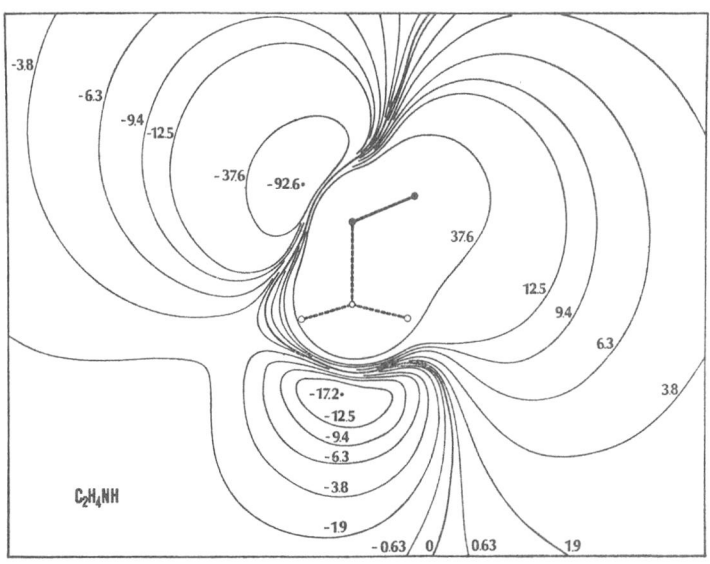

Fig. 9. Potential-energy map for aziridine in the symmetry plane perpendicular to the ring. From Ref. [9]

In the aziridine molecule the shape of the potential is somewhat different. Here too, only two negative regions are found: the first corresponding again to the C—C bond (see Fig. 8) and the second to the heterogroup. The differences between —O— and >NH are, however, well evidenced by the potential. The second negative region is decidedly asymmetric (see Fig. 9) with a sole minimum, in a position again corresponding to the classic direction of the nitrogen lone pair.

From the electrostatic description it follows that in both molecules the most favored reaction channel for an electrophilic reactant X^+ should lead to an intermediate of the type Va, completely different from that suggested for cyclopropane and in complete accord with chemical intuition.

Other cyclopropane derivatives deserve some attention. Oxaziridine (VI) and diaziridine isomers (VII and VIII) are two-substituted derivatives

of cyclopropane and therefore particularly well suited for examining the reciprocal effects of substituents on the potential.

In Fig. 10 two electrostatic potential-energy maps for oxaziridine are reported[12]. Both maps refer to planes perpendicular to the ring and passing respectively through the minimum of $W(\boldsymbol{r})$ adjacent to the NH group and through the two minima pertaining to the O atom. We see how the presence of the oxygen atom reduces by about 13 kcal/mole the depth of the nitrogen hole (with respect to aziridine) and how the presence of the NH group — with the hydrogen atom outside the ring plane — induces noticeable asymmetries in the electrostatic potential near the oxygen (minima of -45 and -32 kcal/mole, respectively, as compared with -46 kcal/mole for the two symmetric minima of oxirane).

The two isomeric forms of diaziridine are different owing to the *cis* (VII) or *trans* (VIII) conformation of the couple of NH groups. Two equivalent minima of $W(\boldsymbol{r})$ are present in both conformers, near the nitrogen atoms, in agreement with what has been found in compounds V and VI. The depth of the holes is, however, greater in the *cis* isomer (-92 kcal/mole) than in the *trans* one (-82 kcal/mole). The difference of 10 kcal/mole must be attributed to the reciprocal influence of the two NH groups: their negative and positive contributions to the potential add

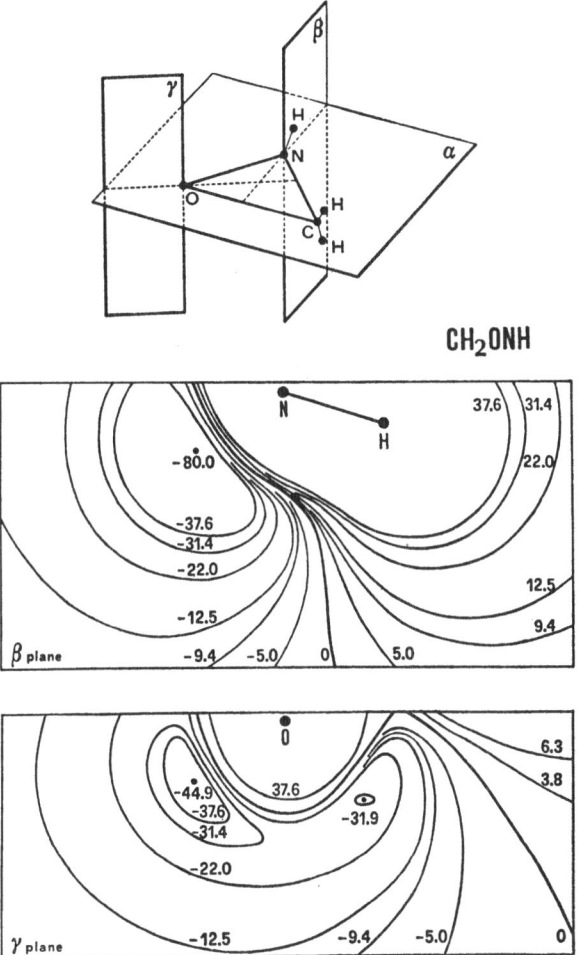

Fig. 10. Potential-energy maps for oxaziridine (VI) in two selected planes. From Ref. [12)]

at least partially, in the *cis* conformer and subtract in the *trans* case. Moreover, this difference is in qualitative agreement with the difference in proton affinity Δ (P.A.)[P.A. $= E$ (AH$^+$) $- E$ (A)] as appreciated on the basis of the SCF energies calculated for A and AH$^+$ using the same basis set employed to get the above-quoted W (**r**) values (Δ (P.A.)$_{SCF} = 7$ kcal/ mole).

C. Unsaturated Compounds

1. Cyclopropene and Derivatives

As examples of molecules containing double bonds, we now consider a couple of compounds structurally related to those of the preceding section, namely cyclopropene (IX) and diazirine (X)

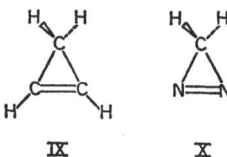

For the first of such molecules an electrostatic potential shape similar to that already found in cyclopropane may be expected, with the exception of the double-bond region where the presence of four electrons in the bent double bond would be expected to enlarge the region where the potential is negative. In fact, in the ring plane (Fig. 11) three negative regions are found, corresponding to the three bonds of the ring. The minima for the single C—C bonds lie in the ring plane, while the map of potential energy in the perpendicular symmetry plane (Fig. 12) shows that the double-bond negative region is decidedly large, with two separated minima. On the whole, the cyclopropene molecule shows four negative holes in the electrostatic potential function, two on the ring plane and two symmetrically placed above and below that plane near the double bond. This situation is in accordance with the description of electronic structure in terms of localized SCF molecular orbitals[9], for the double bond system is represented in this case by a couple of banana bonds bent outwards. The occurrence of the double minimum in the potential is, however, presumably related also to the repulsive contributions of the C—H bonds which are particularly strong in the ring plane and accordingly split into two parts an otherwise unique negative region.

The potential shape in the diazirine molecule is rather different[9]. The remarkable difference in electronegativity between the C and N atoms leads to a notable polarization of the C—N bonds and accordingly no minima have been found in the regions near such bonds (Fig. 13). Moreover, the occurrence of a lone pair for each nitrogen atom with its charge center on the ring plane changes the shape of $W(\mathbf{r})$ in the double-bond region. Fig. 14 — a map in the symmetry plane perpendicular to the ring — shows that the negative region near the double bond is large but without double minima. In conclusion, the diazirine molecule

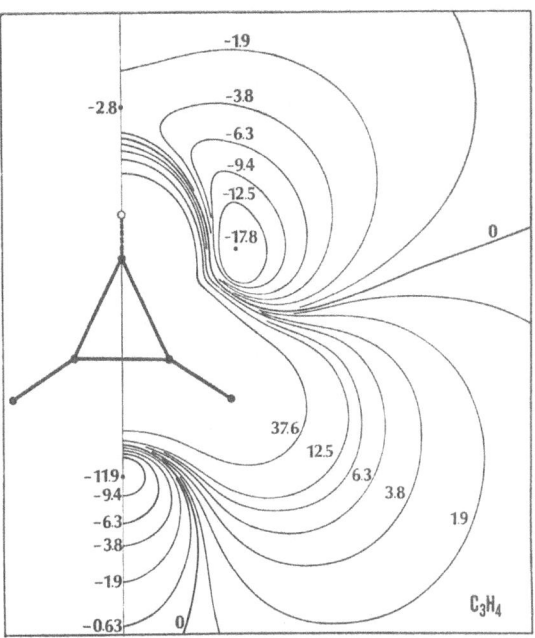

Fig. 11. Potential-energy map for cyclopropene (IX) in the ring plane. From Ref. [9]

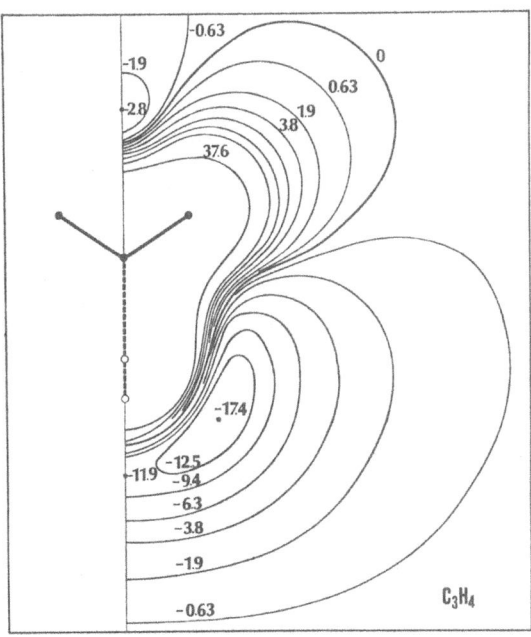

Fig. 12. Potential-energy map for cyclopropene in the symmetry plane perpendicular to the ring. From Ref. [9]

117

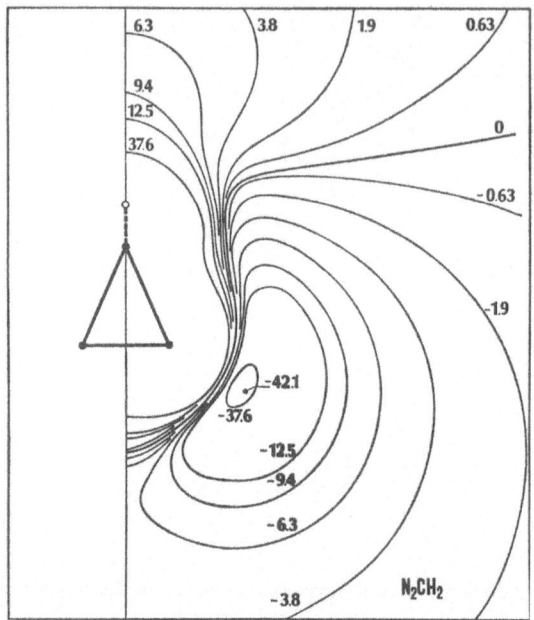

Fig. 13. Potential-energy map for diazirine (X) in the ring plane. From Ref. [9]

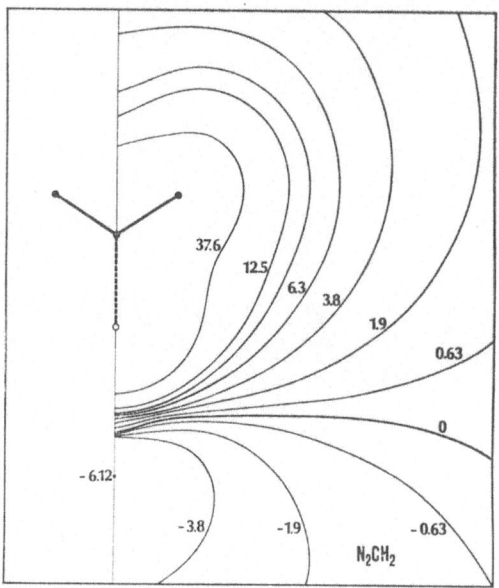

Fig. 14. Potential-energy map for diazirine in the symmetry plane perpendicular to the ring. From Ref. [9]

displays only two holes in the electrostatic potential, corresponding to the N lone pairs whose minima lie in the ring plane (Fig. 13) with a depth (-42 kcal/mole) decidedly less than that found in aziridine and ammonia.

2. Nitrogen Molecule

The examination of the potential near a double bond $-N=N-$ can lead the reader to ask how $W(\mathbf{r})$ behaves near the triple bond $N\equiv N$ in the nitrogen molecule (XI). The map in Fig. 15[5] shows that in the N_2 molecule the bond is surrounded by a positive region. Negative values of $W(\mathbf{r})$ are found only at the ends of the molecule, where two minima, which could be attributed to the two lone pairs, are evident.

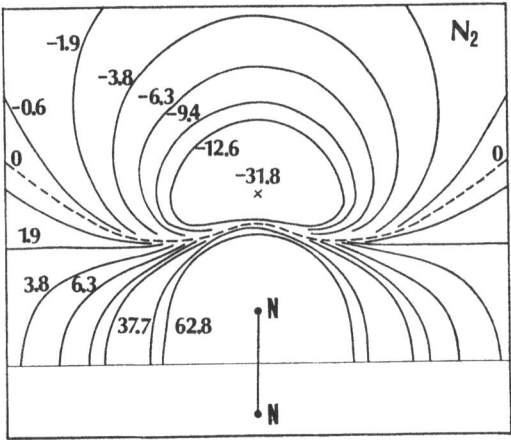

Fig. 15. Potential-energy map for N_2(XI) in a plane containing the nuclei. From Ref. [5]

A comparative examination of all the potential-energy maps for N-containing compounds shows that the N_2 molecule is placed at the lowest level of nucleophility. The difference between N_2 and the other compounds is, however, quantitative and not qualitative. The striking chemical inactivity of N_2 must be tempered if we recall the biochemical fixation of atmospheric nitrogen, of paramount importance in the economy of the biosphere, and the discovery of the coordination compounds of N_2, also obtained under mild conditions[13]. This last category of compounds corresponds to linear coordination, where N_2 acts as either mono- or bidentated ligand. Also the proposed mechanism of fixation (a concerted electron-donor and electron-acceptor action) may be partially explained by an electrostatic picture which involves both negative axial and positive radial regions of N_2.

3. Triatomic Molecules: O_3 and FNO

In order to evidence variations in the molecular potential for isoelectronic molecules of very similar geometry, we consider now the triatomic non-linear molecules, ozone (XII) and nitrosyl fluoride (XIII). Potential maps in the molecular plane are reported for ozone in Fig. 16 and for FNO in Fig. 17[5]. Near the terminal oxygen atoms both molecules

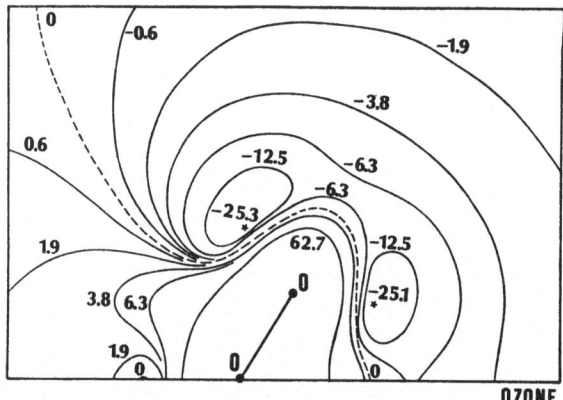

Fig. 16. Potential-energy map for ozone (XII) in the molecular plane

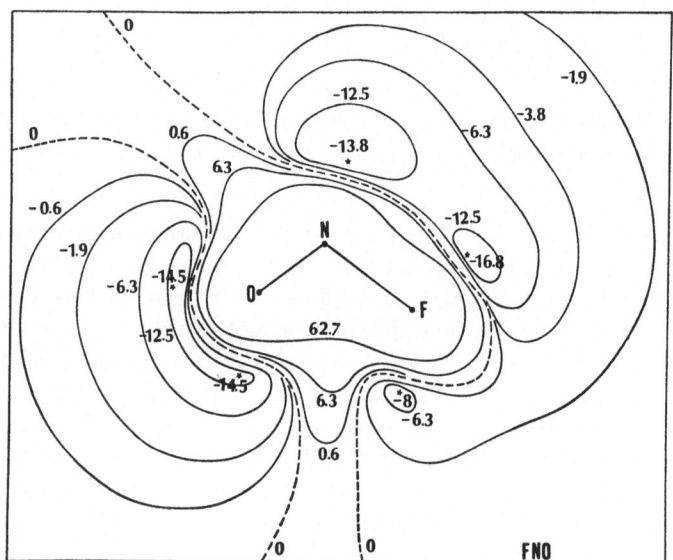

Fig. 17. Potential-energy map for nitrosyl fluoride (XIII) in the molecular plane. From Ref. [5]

display a couple of negative potential holes almost equivalent in shape. The fluorine atom of the second molecule is surrounded by three holes, only one of which lies on the molecular plane.

Near the central atom of ozone there is a flexion in the potential which, however, does not reach negative values, whereas in the corresponding zone of FNO there is an obvious negative hole. The picture given by $W(r)$ is in pleasing accordance with a description of electronic structure in terms of localized orbitals. In ozone, two lone-pair orbitals for each terminal oxygen — with charge center on the molecular plane — and two couples of banana bonds between adjacent oxygen atoms have been obtained[14]. In nitrosyl fluoride one has two lone pairs again for oxygen, a couple of banana bonds between O and N, a single sigma bond between N and F and three lone pairs on the fluorine atom, trigonally projected outwards. In both molecules there is also a lone pair on the central atom, with charge center on the molecular plane. To such lone pairs one could relate the hole found in FNO and the flexion of ozone.

4. Formamide

This molecule (XIV) was selected because it contains two characteristic groups of great chemical importance and because it offers a simple model of the peptide linkage.

The experimental geometry of formamide is practically planar, because of a certain amount of conjugation between the two groups, and therefore the properties of the aminic group in formamide should be different from those of a primary alifatic amine having a pyramidal NH_2 group.

The map of $W(r)$ in the molecular plane is shown in Fig. 18[15]. The carbonyl oxygen is surrounded in the outer region by a wide negative region, as in the preceding examples I, IV, VI, XII and XIII, but in the present case rather than two minima a deep and uniform valley is found[f]. The other portions of the molecule, NH_2 group and C—H bond, are characterized by positive values of $W(r)$.

Fig. 19 shows a map of the electrostatic potential for a plane perpendicular to the molecule and containing N and C atoms. Just above the nitrogen atom there is a small negative region with a minimum of about

---

[f] We note, by the way, that the wave function employed for XIV, as well as for all the following examples was built in terms of Gaussian expansion functions, while the preceding ones were constructed in terms of Slater orbitals. Direct comparisons between the two sets of molecules are postponed unit Section V where some data on basis dependency are discussed.

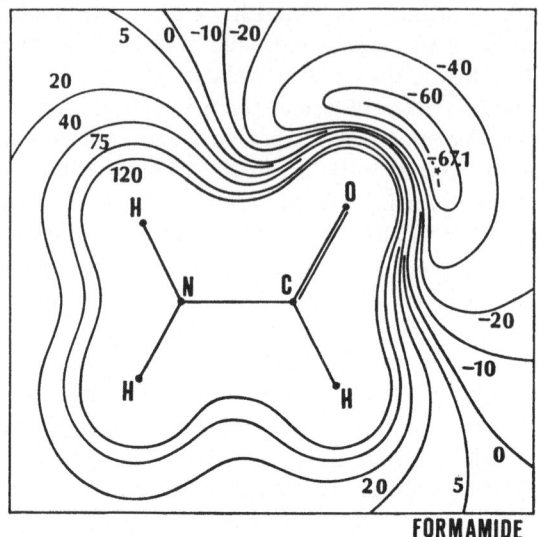

Fig. 18. Potential-energy map for formamide (XIV) in the molecular plane. From Ref. [15]

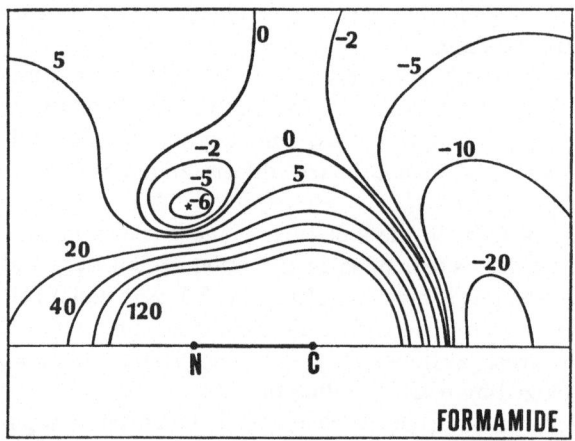

Fig. 19. Potential-energy map for formamide in a perpendicular plane containing N and C atoms. From Ref. [15]

−6 kcal/mole. It is of some interest to note the striking difference in the depth of the hole on passing from pyramidal to plane NH_2 groups: see, *e.g.* II and V. Such differences largely encompass possible differences due to the basis set (see preceding footnote). A moderate pyramidal deformation of the NH_2 group produces considerable intensification of the potential minimum, but is not sufficient to overcome the value of the carbonyl hole[16].

It is clear from this analysis that the most favored approach channel for electrophilic agents is directed towards the carbonyl oxygen. The channel ending on the NH_2 group is decidedly less favoured. As far as the specific reaction of protonation is concerned, this prediction is in complete accord with *"ab initio"* SCF calculations[17]. After a long discussion on the interpretation of relevant experimental data, the experimenters seem to have reached the conclusion at O-protonation occurs first[18].

D. Heteroaromatic Compounds

1. Five-membered Rings

In this paragraph a small series of five-membered heterocycles will be considered. Maps of $W(r)$ in the molecular plane are presented for pyrrole (XV), imidazole (XVI), pyrazole (XVII), oxazole (XVIII) and isoxazole (XIX) in Figs. 20 to 24.

In analogy with the preceding examples, in aromatic compounds too, the electrostatic potential is positive near each hydrogen atom, while negative regions are present near the $-N=$ and $-O-$ atoms. One can observe a very small negative region in pyrrole corresponding to the C_3-C_4 bond. This electrostatic picture correctly shows that $-N=$ heteroatoms are more reactive towards electrophilic agents than $-O-$ atoms. From the point of view of correlations, we note that the ordering of proton affinities in XVI, XVII, XVIII and XIX, as measured by the corresponding pK_a's (6.95[19], 2.48[20], 0.8[21], -2.03[21], respectively) is the same as that of $W(r)$ minima[22].

A correct forecast of differences in electrophilic reactivity among the carbon atoms of the ring constitutes a classical testing bench for all theories on chemical reactivity. To exploit the electrostatic approximation, we need some information on the trend of $W(r)$ in the regions

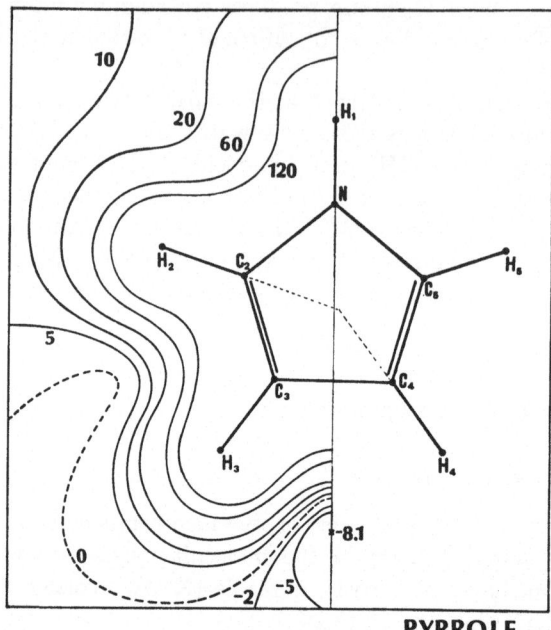

Fig. 20. Potential-energy map for pyrrole (XV) in the ring plane

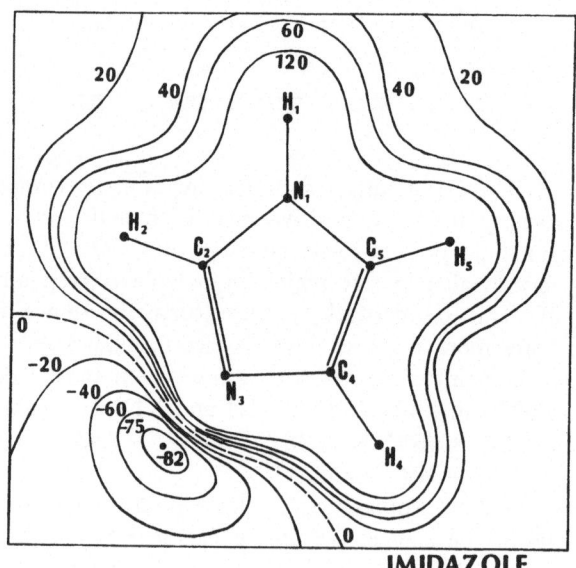

Fig. 21. Potential-energy map for imidazole (XVI) in the ring plane

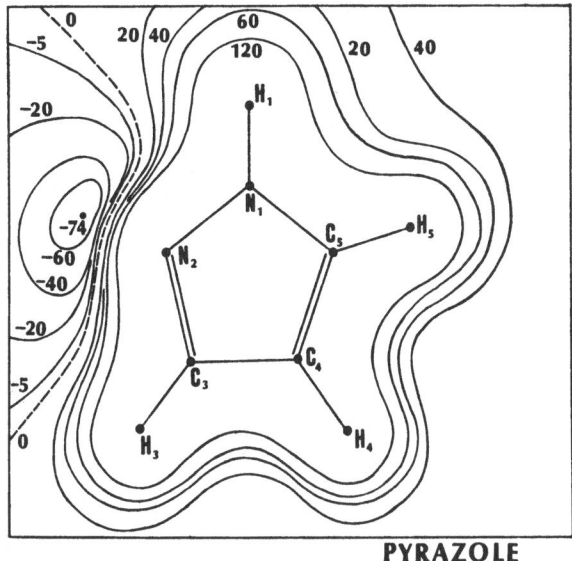

Fig. 22. Potential-energy map for pyrazole (XVII) in the ring plane

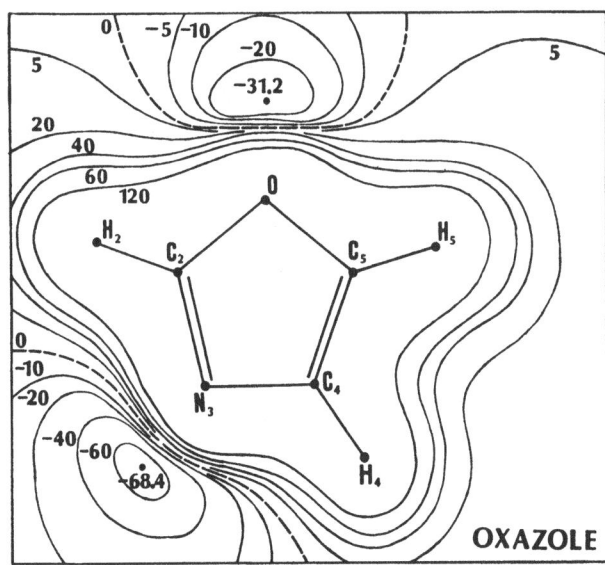

Fig. 23. Potential-energy map for oxazole (XVIII) in the ring plane

125

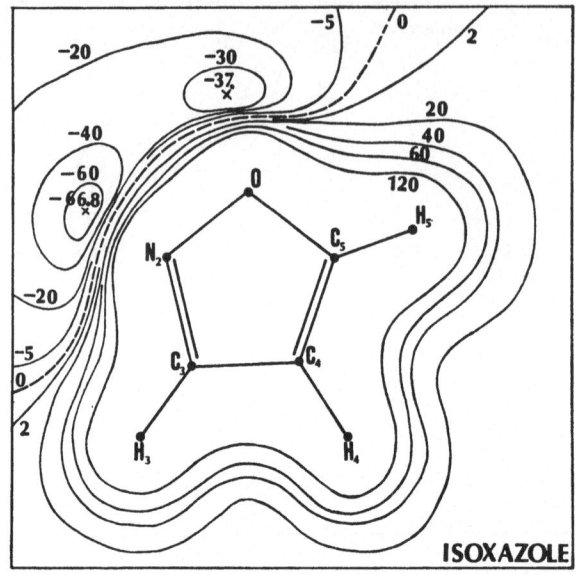

Fig. 24. Potential-energy map for isoxazole (XIX) in the ring plane

outside the ring plane. In the series of Figs. 25 to 32 we have reported sufficient material to test the electrostatic predictions. These figures give maps of $W(r)$ drawn for planes perpendicular to the ring and containing C—H nuclei. The identification is given in the captions. A first glance at these figures shows that the electrostatic potential, positive near the ring plane, becomes negative (*i.e.* attractive for electrophiles) at larger distances.

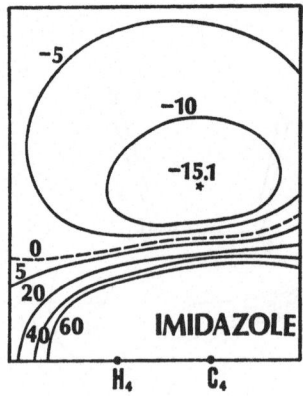

Fig. 25. Potential-energy map for imidazole in a plane perpendicular to the ring and containing C_4, H_4 atoms

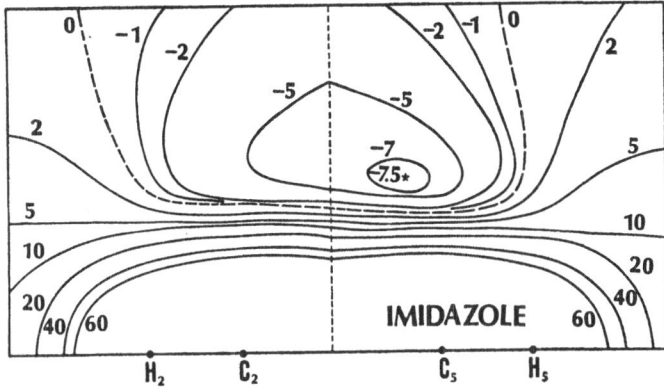

Fig. 26. Potential-energy map for imidazole in two perpendicular planes containing H_2, C_2 and C_5, H_5, respectively

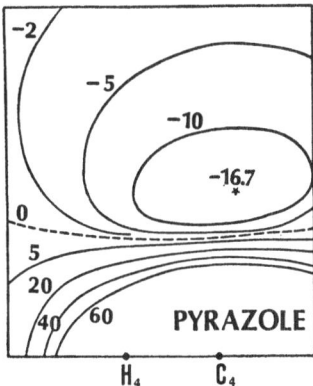

Fig. 27. Potential-energy map for pyrazole in a plane perpendicular to the ring and containing H_4 and C_4 atoms

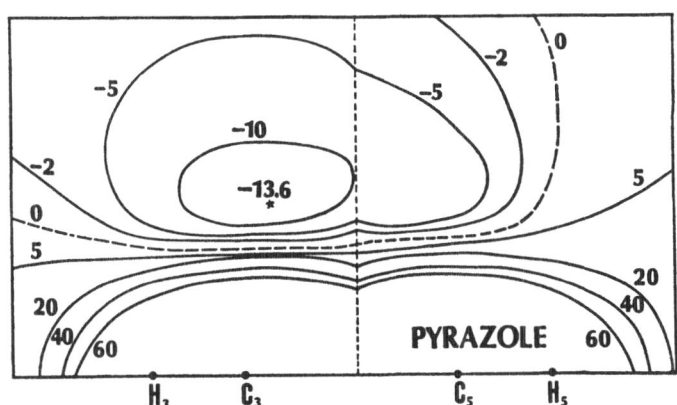

Fig. 28. Potential-energy map for pyrazole in two planes perpendicular to the ring and containing H_3, C_3 and C_5, H_5 atoms, respectively

127

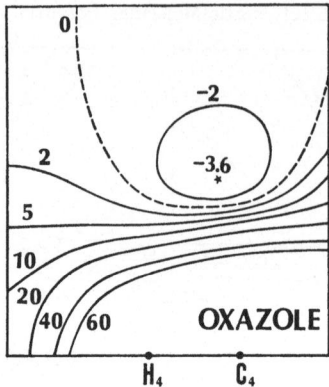

Fig. 29. Potential-energy map for oxazole in a plane perpendicular to the ring and containing H_4 and C_4 atoms

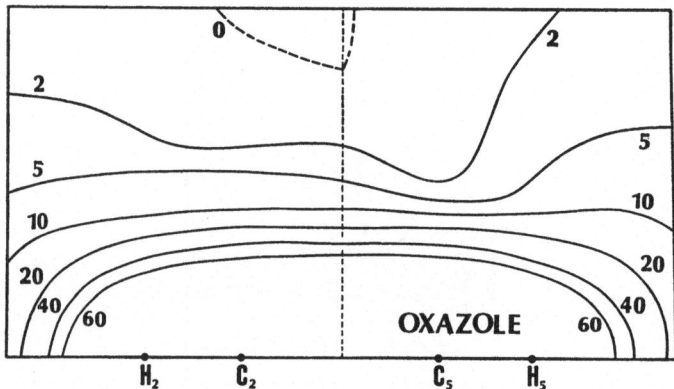

Fig. 30. Potential-energy map for oxazole in two planes perpendicular to the ring and containing H_2, C_2 and C_5, H_5 atoms, respectively

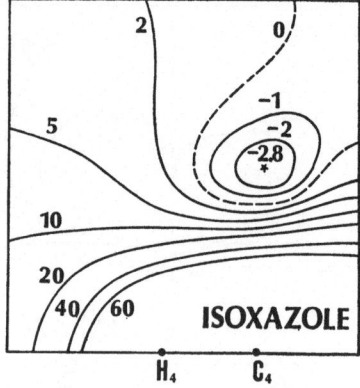

Fig. 31. Potential-energy map for isoxazole in a perpendicular plane containing H_4 and C_4 atoms

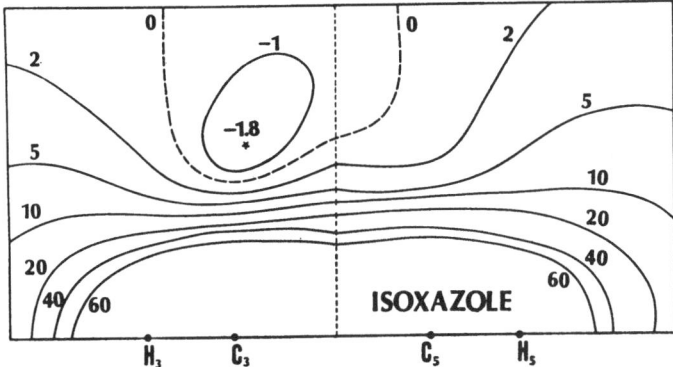

Fig. 32. Potential-energy map for isoxazole in two planes perpendicular to the ring and containing H_3, C_3 and C_5, H_5 atoms, respectively

Electrostatically, the reaction channels reach the carbon atoms through the π region of the ring. On a closer examination, clear-cut differences among the different positions become evident. In the four compounds containing two heteroatoms, position 4 has in all cases the deepest hole and is accordingly the most reactive towards electrophilic attacks. The compounds containing two nitrogen atoms (XVI and XVII) show, in addition, deeper channels than compounds XVIII and XIX containing nitrogen and oxygen. The electrostatic characterization of reactivity of the carbon atoms in these four compounds seems not to be in sharp

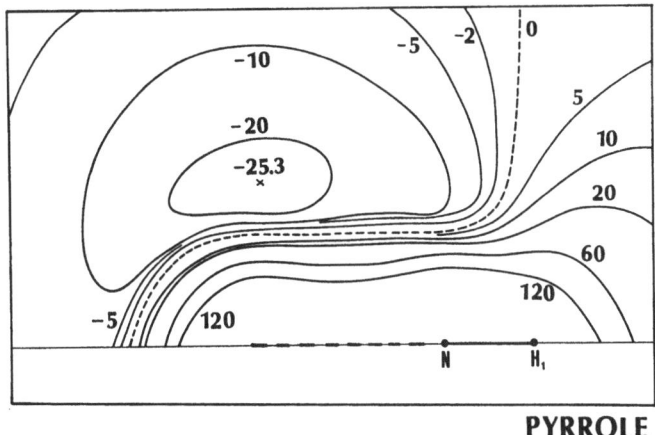

Fig. 33. Potential-energy map for pyrrole in the symmetry plane perpendicular to the ring

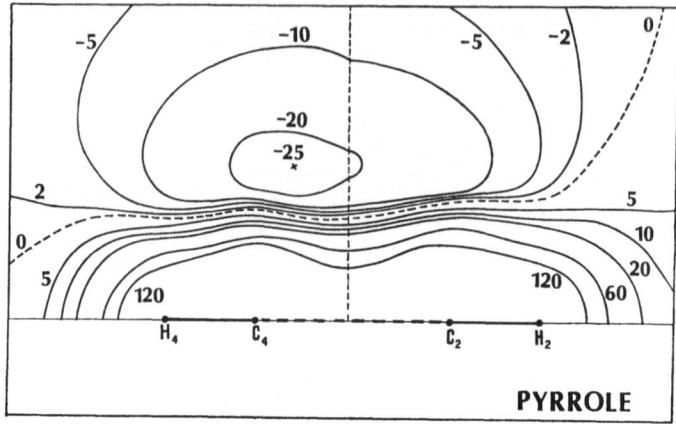

Fig. 34. Potential-energy map for pyrrole in two planes perpendicular to the ring and containing H_4, C_4 and C_2, H_2 atoms, repectively

contrast with chemical plausibility and agrees with the few experimental data related to neutral molecules[23].

In pyrrole (XV), a wide approach channel for electrophilic reagents leads to positions 3 and 4 (see Figs. 33 and 34). This finding is in accordance with the experimental evidence[24] that protonation in 3,4 is faster than in 2,5, though the 2-protonated salts are more stable. As has been repeated many times, the electrostatic approximation can give at most a picture of the first part of the reaction and it is not able to predict the energetically most stable final product.

2. Six-membered Heterocycles

We have at present electrostatic potential maps only for pyridine (XX) and pyrazine (XXI)

XX   XXI

In both compounds large negative regions surround the heteroatoms, in analogy with what has been found in the five-membered heterocycles. The minimum for pyridine (-68.3 kcal/mole) is deeper than that of pyrazine (-61 kcal/mole), in accordance with the greater basicity of the

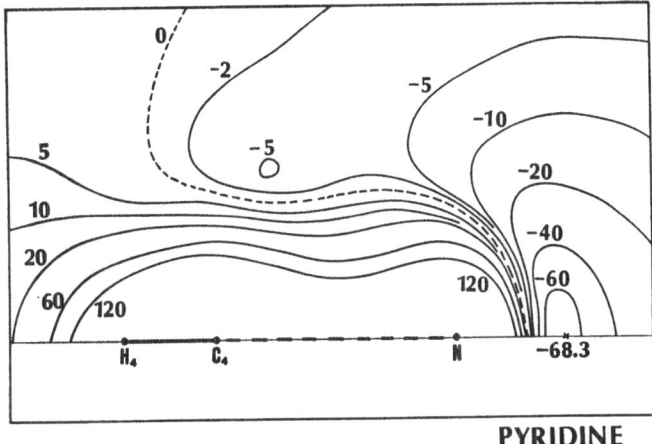

Fig. 35. Potential-energy map for pyridine (XX) in the symmetry plane perpendicular to the ring

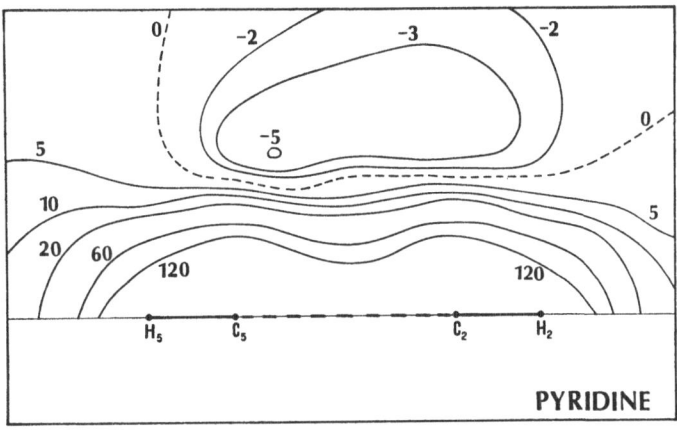

Fig. 36. Potential-energy map for pyridine in a plane perpendicular to the ring containing C_2 and C_5 nuclei

former ($pK_a = 5.23$ for XX[25)] and 0.65 for XXI[26)]). The correlation line parallels that found for compounds XVI-XIX. It is necessary to consider that in order to find a correlation among free energies for molecules in solution (as deduced from pK_a's) and gas-phase enthalpies (as approximated by $W(r)$ values), other quantities — like the changes in solvation energy following protonation and the entropy variations — must remain constant throughout the set of molecules considered. It is questionable

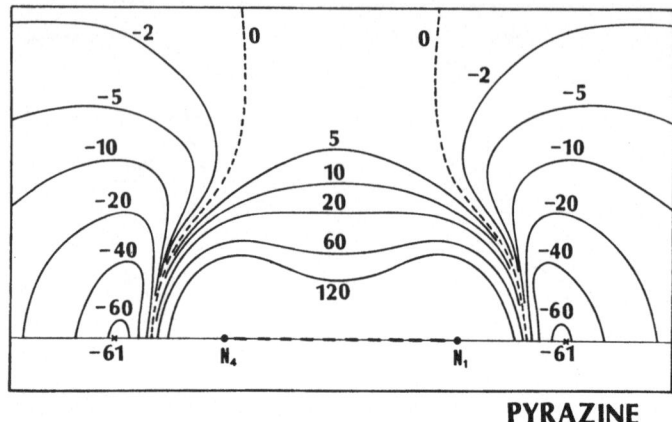

Fig. 37. Potential-energy map for pyrazine (XXI) in the symmetry plane perpendicular to the ring

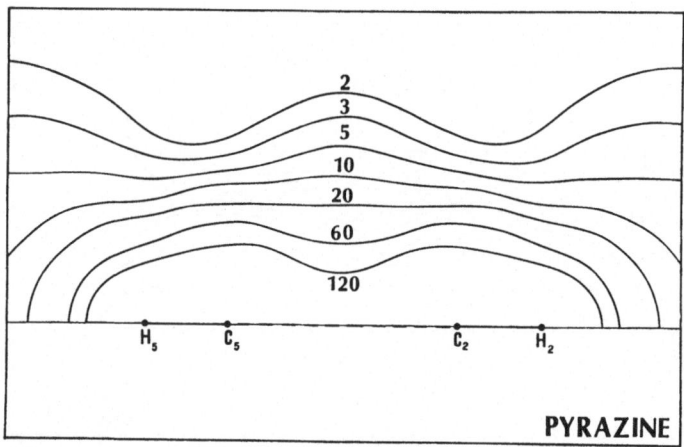

Fig. 38. Potential-energy map for pyrazine in the plane perpendicular to the ring containing C_2 and C_5 atoms

whether one may assume that such variations are negligible over the set XVI-XXI.

The electrostatic potential of pyridine outside the ring plane presents large negative regions with small minima (nearly equivalent) in correspondence with atoms C_3-C_5 and C_4. It is not easy to find clear-cut examples of reactions involving the free base. Indirect evidence indicates that atoms C_3-C_5 are the most reactive. Nucleophilic substitution reactions, which are generally found to be favored in positions 2-6, could indirectly suggest that the C_4 atom is more electronegative than C_2-C_6, but this inference has been criticized[27].

In the pyrazine molecule the molecular potential outside the ring plane shows some peculiarities: the negative region does not spread over the whole ring but is broken near the two N atoms (Fig. 37). The large electron attraction effects of the two nitrogen atoms partially deshield the carbon nuclei, and the behavior of $W(r)$ offers an explanation of the greater resistance towards electrophilic attacks of pyrazine compared with pyridine. The experimental reactivity of the single type of carbon atoms in pyrazine is claimed to be comparable to that of C_2 atom in pyridine[28]. Such a qualitative comparison is not well accounted for by the electrostatic potentials, though in neither case does electrophilic attack seem particularly favored.

3. Purinic Bases

As a last example of shape of electrostatic potential, we report here some results for a purinic compound, adenine (XXII). Analogous results for other nucleic acid components, thymine and cytosine, may be found in the source paper[29].

XXII

In analogy with the preceding examples, three negative zones, corresponding to the three =N-atom, are found, with minima in the plane of the rings (Fig. 39). Other two symmetric minima, of decidedly lower value, are found above and below the $-NH_2$ group (Fig. 40). The basicity of pyridine-like nitrogens is much larger than that of the aminic nitrogen, in accord with the interpretation of experimental facts now generally accepted. The minima corresponding to N_1 and N_3 are practically equivalent and slightly deeper than the one corresponding to N_7. Experimentally, the preferred positions for an electrophilic attack were found to be N_1 and N_3 (N_1 for protonation[30], N_3 for alkylation[31]).

It may be of some interest to note that a remarkable analogy in form and values of $W(r)$ was found between the out-of-plane region of imidazole and the corresponding region of the imidazole moiety of

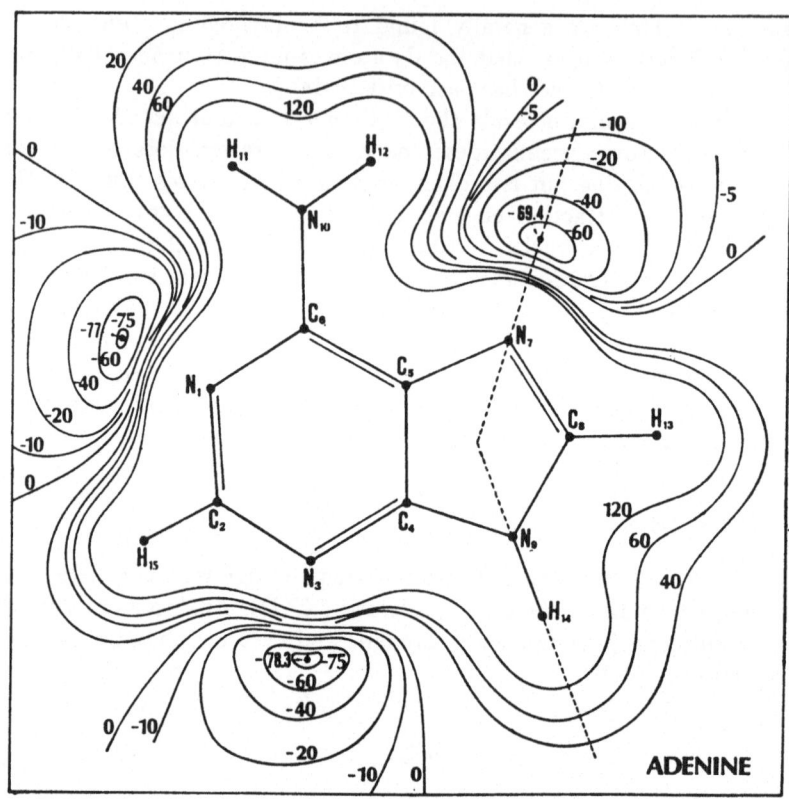

Fig. 39. Potential-energy map for adenine (XXII) in the ring plane. From Ref. [29]

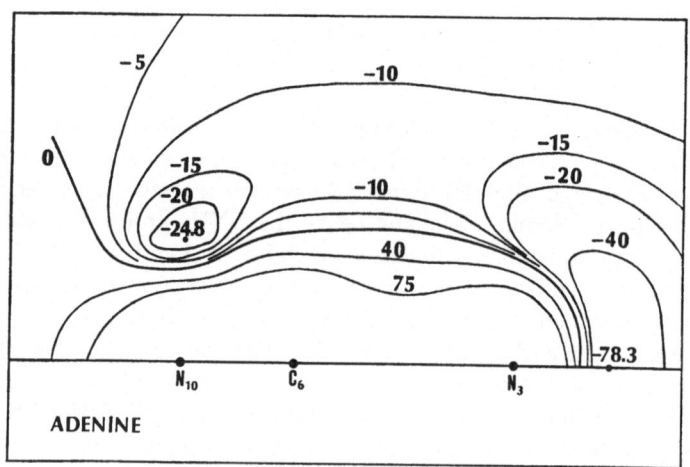

Fig. 40. Potential-energy map for adenine in a plane perpendicular to the ring and containing atoms N_{10}, C_6 and N_3. From Ref. [29]

adenine (no maps are given here). This kind of finding suggests that one could investigate the conservation properties of the potential relative to a molecular fragment inserted into different chemical frameworks. Such topics will be considered in Section VII.

E. Provisional Conclusions

The examples of electrostatic molecular potential reported above are very far from being sufficient to permit general conclusions to be drawn. It is however, convenient to place here some provisional observations:

a) The C—H, N—H, O—H bonds are characterized by positive regions of the potential. In most cases they constitute a sort of wall for the electrophilic reaction channels.

b) For the N and O heteroatoms, which in the classical description are provided with electron lone pairs, the electrostatic potential shows well-evidenced negative holes.

c) The potential hole is generally deeper for N atoms than for O atoms.

d) In a given chemical family, it seems possible to get linear correlations between the depth of the hole and the basicity of the corresponding chemical position.

e) The following sequence is found for the depth of $W(\mathbf{r})$ holes for nitrogens in conjugated compounds: pyridine-like $(\geqslant N) >$ planar trigonal $(-NH_2) >$ pyrrole-like $(>N-H)$.

f) In aromatic compounds negative values are found also above and below the ring, and the characteristics of such zones seem to be potentially related to the reactivity toward electrophilic reagents of the different atoms of the ring.

g) The trend of $W(\mathbf{r})$ in the above-mentioned regions is not always correlated with π charges. A better correlation is found with total *ab initio* charges.

V. The Dependence of $V(r)$ on the Accuracy of the Wave Function

A. SCF Wave Functions

An important question related to the reliability of the results of the preceding section is how much they depend on the accuracy of the wave functions employed.

As already noted, all the examples in the preceding section refer to SCF ab initio wave functions, which do not take into account electron correlation. It is well known, however, that the SCF approximation at the Hartree-Fock limit is good enough to give a reliable representation of a one-electron, first-order observable like the electronic potential, at least for closed-shell ground-state systems like those considered here. If we keep to the field of SCF *"ab initio"* wave functions, the differences in accuracy between several wave functions for the same molecule depend upon the adequacy of the expansion basis set χ employed in the calculations. In the next paragraph, however, we will also treat the case of semiempirical SCF wave functions.

The examples quoted in Section IV all refer to minimal basis set wave functions composed of Slater-type orbitals (best atom zetas[32]) for molecules I-XIV, and of Gaussian orbitals for the others. Wave functions for compounds XV-XXI refer to a (7s 3p|3s) basis contracted to [2s 1p|1s] proposed by Clementi, Clementi and Davis[33] (CD basis), while for compound XXII we have used another Gaussian basis (4s 2p|3s) contracted to [2s 1p|2s] proposed by Mély and Pullman (MP basis)[34].

For some molecules electrostatic potentials calculated with other wave functions are available. We can anticipate that the essential features of the shape of the electrostatic potential are conserved when one changes SCF wave function. Remarkable variations in the absolute values of $W(r)$ are on the contrary observed.

For H_2O, with a rather extended basis[35] 29 Slater-type orbitals, including polarization d orbitals on oxygen and p orbitals on hydrogen-one obtains a very similar division into positive and negative regions. The position of the minima changes slightly — they are a little more separated — but their value is −49 kcal/mole instead of the 73.7 kcal/mole found with BAZ wave function. The optimization of orbital exponents in the molecule (minimal basis set) produces small changes in the value of the minima: Ref.[36] gives the map of $W(r)$ for water calculated with the STO basis optimized in the molecule by Aung, Pitzer and Chang[37]. The value of $W(r)$ at the minima is −79.6 kcal/mole. With the CD Gaussian basis, it has practically the same shape as in Figs. 2 and 3, and minimum values of −70.3 kcal/mole[38]. Ref.[36] also contains $W(r)$ maps for formaldehyde with BAZ STO's as well as with BMZ's (best molecular zeta) STO's (in both cases minimal basis sets). Near the oxygen two minima have been found: −35 kcal/mole in the first case and −47.6 in the other. Formaldehyde maps are also available on a Gaussian CD basis: the minima are at −34.6 kcal/mole[38]. A comparison between the two Gaussian bases CD and MP performed in formamide is reported in Ref.[15]. Some maps corresponding to the two wave functions are placed side by side to make comparisons easier: the shape of $W(\mathbf{r})$ is very similar and differences among the two sets of results are about 4 to 8 kcal/mole in the minimum regions, the CD values being lower.

Rather than quote similar results on another scattered series of molecules, we pass to a more quantitative examination on a chemical family.

A comparison has been made[39] among the electrostatic potentials of three-membered ring molecules III-X (and others not reported in the present paper) calculated with STO BAZ wave functions (*i.e.* those reported in Section IV) and with CD Gaussian wave functions. The number of negative zones and the position of the minima show sufficient agreement. Also in this set, absolute values of $W(\mathbf{r})$ are rather different. In Fig. 41 we compare the values of the same minima obtained *via* the

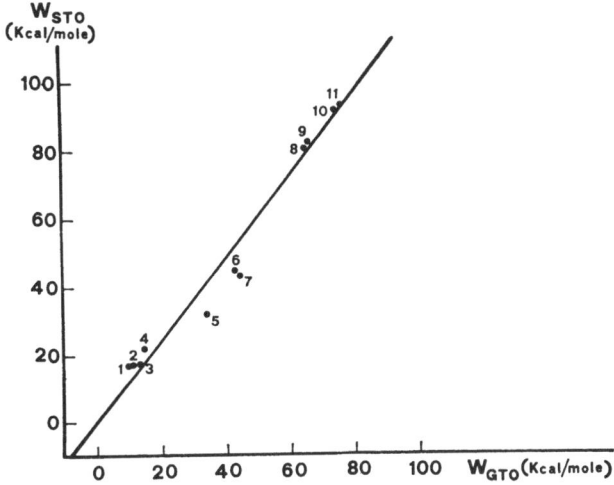

Fig. 41. Comparison of the values of minima of electrostatic potential in some three-membered ring molecules according to STO and GTO minimal basis set SCF wave functions. The values are labeled as follows: 1, aziridine (C—C), 2, cyclopropene (C=C), 3, cyclopropene (C—C), 4, cyclopropane (C—C), 5, oxaziridine (O), 6, oxaziridine (O), 7, oxirane (O), 8, oxaziridine (N), 9, trans-diaziridine (N), 10, cis-diaziridine (N), 11, aziridine (N)

two different bases. A fairly linear correlation between the two sets of results is evident (the correlation coefficient is 0.984).

The provisional conclusion reached on the basis of checks performed so far is that, for molecules built up with atoms of the first and second row of the periodic table, a minimal basis set SCF wave function is good enough to reveal the outstanding features of the potential at a level sufficient for qualitative identification of chemical sites. Semi-quantitative comparisons are, of course, feasible only if the interested wave functions are built up in terms of the same basis.

B. Semi-empirical Wave Functions

For a large-scale application of electrostatic potentials — comparisons among large sets of molecules and investigations on big molecules — it would be desirable to be able to resort to semi-empirical wave functions, which can be computed a good deal faster than *ab initio* SCF ones. It is necessary, however, to ensure that the reliability of the $W(r)$ values is not too much affected by going over to approximate wave functions.

Preliminary investigations on the reliability of semiempirical $W(r)$ maps have been started. They are related to the CNDO/2 method. Some methods of deriving a $W(r)$ from a CNDO wave function have been examined by Giessner-Prettre and Pullman[36]. Among the several approximations they have considered, the most involved (called in Ref.[36] approximation IV) gives the best results. It relies on a transformation of the CNDO MO coefficients, considered as corresponding to a Löwdin orthogonalized Slater-type basis set ν, into other coefficients corresponding to a normal (not orthogonal) Slater-type basis set χ:

$$c^\chi = S^{-\frac{1}{2}} c^\nu$$

where S is the overlap matrix written in terms of χ functions: $S_{ij} = \langle \chi_i | \chi_j \rangle$. After performing such a transformation, if one takes into account explicitly the contributions to electrostatic potential arising from all the two-center distributions $\chi_i \chi_j$, one arrives at $W(r)$ maps roughly comparable with the *ab initio* ones.

As regards the absolute values of the electrostatic potential, we report the minima for H_2O: -74.2 kcal/mole and for H_2CO: -55.5 kcal/mole obtained by Giessner-Prettre and Pullman[36] with the above-defined approximation IV.

Another comparison is given in Fig. 42[40]. It refers again to the three-membered cycles III-X. The minima found with CNDO wave functions are compared with those arising from CD Gaussian wave functions[39]. The correlation coefficient is 0.978. An analysis of the shape of $W(r)$ for such molecules in the CNDO approximation shows, however, that the secondary minima near the C–C bonds in heterocycles tend to disappear. CNDO wave functions in fact give a charge transfer from carbon to a heteroatom larger than the *ab initio* SCF wave functions,

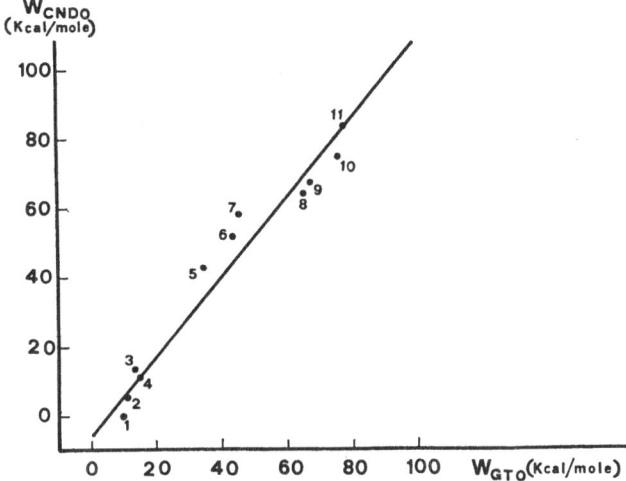

Fig. 42. Comparison of the values of minima of electrostatic potential in some three-membered ring molecules according to GTO minimal basis set SCF wave functions and to CNDO semiempirical calculations. The values are labeled as in the preceding figure

on the Slater basis[9,12] as well as on the Gaussian basis[39]. Therefore some caution must be observed in using semiempirical potentials, especially when passing to new families of compounds. At the present state of our knowledge about this subject, the use of semiempirical potentials looks promising, although more extensive checks are needed.

A straight forward application of approximation IV to calculate $W(r)$ maps is quite exacting, because the calculation of the potential contribution due to the couple distributions $\chi_i \chi_j$ is time consuming when directly performed on the Slater functions. This fact clashes with the basic philosophy of semiempirical methods, which is to sacrifice some reliability to speed up the calculations. It has been shown[40] that expansion of each Slater-type orbital into three Gaussian functions (3 G expansion[41]) gives a substantial improvement of the computational times of $W(r)$, without an appreciable reduction in the quality of the results.

VI. Protonation Processes

The shape of the electrostatic potentials was employed in Section IV to reveal the reactivity capabilities (position and depths of the approach channels for electrophilic reagents, values of the $W(r)$ minima, etc.)

of the single chemical groups constituting the molecule, with tacit acceptance of the electrostatic assumption. But if we consider an actual case from a quantitative point of view, the electrostatic approximation can raise grave doubts. For example, the maximum electrostatic inter- action energy of water with a proton is about −80 kcal/mole (with a minimal basis set SCF wave function) while experimentally the protona- tion energy, in gas phase, is about −180 kcal/mole[42]. In general the $W(r)$ values are less than one half of the correct value of protonation.

Admittedly, interaction with a bare proton is a limiting case, but it is expedient to analyze a limiting case before going over to actual utiliza- tions.

In the case of the protonation process of a molecule A, $i.e.$ the ap- proach of a proton to A, the large field arising from the proton itself generates appreciable polarization in the electron cloud of A. Charge transfer to the proton will also be of some importance at medium dis- tances. We should point out however, that the electrostatic approxima- tion "$per\ se$" can tolerate quite large differences between electrostatic and correct interaction energies: the essential point is that the ratio between these values must remain constant for some portions of the space. Because many of the applications of the electrostatic potential are comparative in character, it is usually sufficient to have a functional dependence between the interaction energy and its electrostatic part. Such functional dependence must, of course, remain constant among the set of molecules which form the subject of the comparison.

To check the electrostatic picture of the protonation processes, we return again to the three-membered ring molecules. The reader will recall that on electrostatic grounds the prefered positions of primary protona- tion were found near the heteroatom lone pairs and the bent C−C bonds of the rings. Fig. 43 is a graphical comparison between $W(r)$ values corresponding to the positions of the minima, and SCF protonation energies $\Delta E(r) = E(AH^+) - E(A)$ (calculated at the same positions). $E(A)$ is the energy of the wave function used to calculate $W(r)$ (in Fig. 43 the values refer to the CD Gaussian basis wave functions[39]), while $E(AH^+)$ is the energy of a protonated species having the same geometry as A with the extra proton placed at the same position as the correspond- ing $W(r)$ minimum[g].

In the chemical family here considered, a linear relationship between $W(r)$ minima and SCF values of protonation energy (calculated at the same positions) is evident from Fig. 43; the correlation coefficient of the

---

[g] For the AH⁺ species the same CD basis set was used as for A, and the same subset of atomic orbitals already employed for the other hydrogens of the molecule was assigned to the proton.

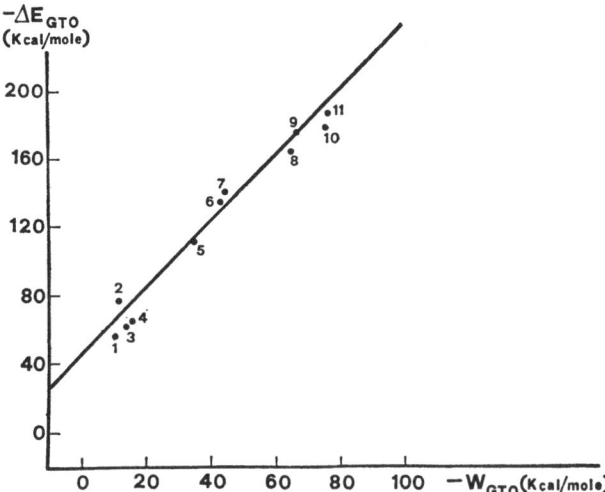

Fig. 43. Comparison between the SCF proton interaction energy (ΔE_{SCF}) and the value of the electrostatic potential at the minima for some three-membered ring molecules. Both sets of calculations refer to Gaussian basis (CD) wave functions. The points are labeled as in Fig. 41

regression line is 0.988. Therefore, from the $W(r)$ values one can obtain at least an ordering of the various primary protonation processes.

Another important aspect of the same problem is the reliability of the electrostatic representation of the approach channels. Fig. 44 depicts the energy situation for the path of most direct approach for a proton to the N lone pair of aziridine (V): practically a straight line passing through the N nucleus and the $W(r)$ minimum. Curve a) of Fig. 44 gives the $W(r)$ values along this trajectory, curve b) represents the sum of both the electrostatic and polarization energies (molecular Hartree approximation, see Section II. D), and curve c) gives the interaction energy ΔE calculated as the difference between the SCF energy of AH⁺ — calculated for the different positions which the proton takes up along the path — and that of A. Curves a) and b) both characterize in a qualitatively reasonable manner the energy trend along these approach channels. In particular, the minima of the three curves lie very close together (1.20 Å for a, 1.06 Å for b and 1.12 Å for c). A perpendicular section of the same channel (at 2.75 Å from the nitrogen nucleus) is reported in Fig. 45. The values of Figs. 44 and 45 refer to the CD Gaussian wave functions[39].

Similar results have been obtained for other molecules (NH_3[43], H_2O[43], H_2NCO[17a], etc.) on minimal basis set functions. By removing the constraint of contraction factors in the CD basis one obtains equivalent results. No controls have yet been performed on wave functions

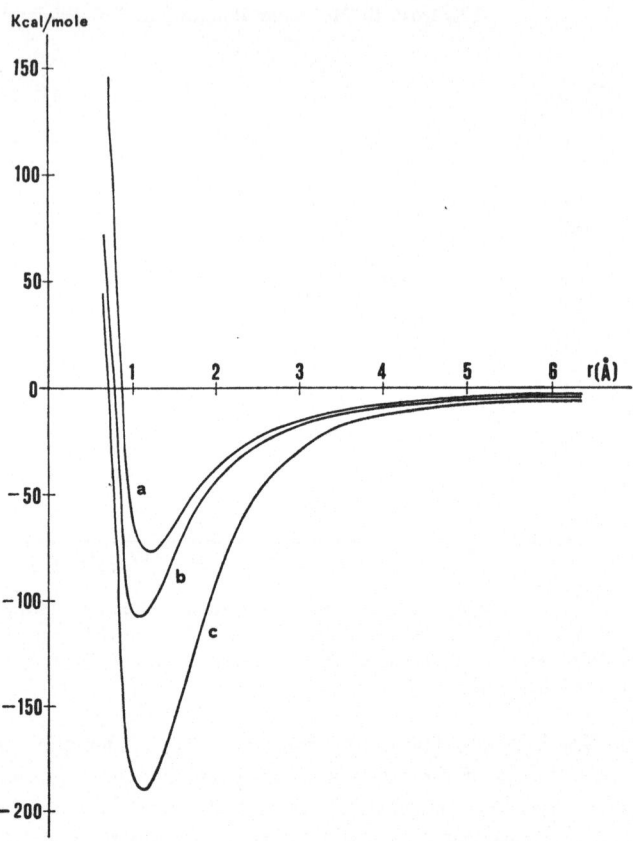

Fig. 44. Proton interaction energy trend along an approach path to the aziridine N, as obtained a) by the electrostatic approximation, b) by the Hartree approximation, c) by SCF computations (GTO wave function)

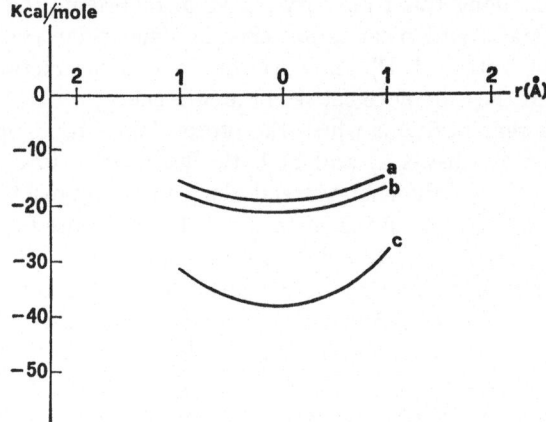

Fig. 45. A section, at $R = 2.75$ Å, of the proton approach channel shown in Fig. 44

closer to the H−F limit. When semiempirical CNDO representations of the electrostatic potential are examined, one reaches analogous conclusions concerning the correlation between ΔE and $W(\mathbf{r})$ as well as the electrostatic description of the shape of the channels. Such results, however, are only provisional because the checks so far performed concern only the set of three-membered rings[40] and the formamide molecule[44].

VII. Group Contributions to the Electrostatic Potential

One of the main problems in structural chemistry is the attempt to divide molecular properties (like dipole moment, electric polarizability, formation enthalpies, reactivity, etc.) into contributions arising from the various subdivisions of the molecule. The basic assumption here is the *conservation* of the properties of each single chemical group and its (approximate) *transferability* from one molecule to another. In this section we sketch a brief approach to the specific case of the group contributions to $V(\mathbf{r})$. An operative definition of group potential will be followed by an examination of the relative importance of both near and far groups to the electrostatic potential at a given point, and finally by a verification of the degree of conservation of group potentials.

A. Localized Orbitals and Related Partition of $V(r)$

The SCF wave functions we have used to calculate $V(\mathbf{r})$ are written in terms of *canonical* one-electron orbitals φ_i which spread over the whole molecule. Canonical form is not able to give a simple and evident visualization of a single bond or chemical group. A better representation of the wave function for this purpose is in terms of *localized* orbitals (LO's), which give a chemically more expressive picture of the electron distribution. It is well known that a one-determinant wave function, written in terms of canonical orbitals φ_i, can be transformed into another, completely equivalent one, written in terms of localized orbitals λ_i. It is merely necessary to perform a suitable unitary transformation on the set φ:

$$\lambda = \varphi U$$

Various methods of constructing a unitary matrix U with localizing properties have been proposed [for a review, see Ref.[45]]. For the present analysis we have adopted Boys' method[46], as being the simplest intrinsic method (for a definition of intrinsic versus external methods, see Ruedenberg[47]].

143

The invariance of the first-order density matrix with respect to unitary transformations ensures the invariance of all one-electron properties, like electrostatic potentials. Thus the transformation to localized orbitals does not alter the value of the potential at any point r of the space, but permits a chemically meaningful partition of this quantity. In fact, the "lone pair", "bond" and "core" localized orbitals resulting from the Boys' transformation are particularly suitable for our attempt a) to give a rational basis to the additivity rules for group contributions, and b) to find some criteria by which to measure the degree of conservation of group properties.

Analyses of this kind have already been performed on other properties[48]; here we discuss only the specific case of $V(r)$. The chemical motivation for such an analysis lies in the hope of getting a qualitative interpretation of reactivity variations among the molecules of a given family. Such an interpretation will certainly be incomplete and limited to reactions for which it is possible assume an electrostatic mechanism as a first order approximation.

If the definition (11) of the electronic charge density distribution is cast in terms of localized orbitals λ_i, one obtains for the total molecular charge distribution (Eq. 10):

$$\gamma(r_1, R) = -2\sum_i \lambda_i^*(r_1)\lambda_i(r_1) + \sum_a Z_a \delta(r_1 - R_a) \qquad (14)$$

and for the electrostatic molecular potential:

$$V_A(r) = -2\sum_i \int \frac{\lambda_i^*(r_1)\lambda_i(r_1)}{|r - r_1|}\, dr_1 + \sum_a \frac{Z_a}{|r - R_a|}. \qquad (15)$$

Definition (15) evidences the additivity of the electronic contributions of each localized pair, but it does not properly associate to every such contribution a corresponding nuclear part. It is convenient, therefore, to rewrite the nuclear contributions in another form, but divided according to the same scheme.

The partition we have adopted associates to each electron pair two unit positive charges selected in the following way:

a) If the localized orbital λ_i corresponds to a bond orbital between two atoms α and β, one positive charge is assigned to the α nucleus and the other to the β nucleus.

b) If λ_i corresponds to a core atomic orbital, or to a lone pair of the valence shell, both positive charges are assigned to the pertinent nucleus.

144

Such a partition, which preserves the electroneutrality in each molecular fragment, is not unusual in intuitive arguments on molecular structure and has already been employed for similar analyses of chemical properties in terms of localized orbitals[9,29]. The electrostatic potential may therefore be written as:

$$V(r) = \sum_i \left[-2 \int \frac{\lambda_i^*(r_1)\,\lambda_i(r_1)}{|r - r_1|}\, dr_1 + \frac{1}{|r - R_\alpha^{(i)}|} + \frac{1}{|r - R_\beta^{(i)}|} \right] \quad (16)$$

where $R_\alpha^{(i)} = R_\beta^{(i)}$ if the localized orbital λ_i is of core or lone-pair type. It is evident that (16) may readily be extended to larger groups, like the conventional functional groups (CH_3, CH_2, NH_2 etc.).

B. An Example of Analysis of $W(r)$

The saturated three-membered ring compounds of Section IV. B. 2 will be used here again to exemplify the analysis of $W(r)$. Such cycles constitute, in fact, the largest available set of molecules with different groups inserted in the same rigid molecular framework.

Table 1 shows the partition of $W(r)$ for the oxaziridine (VI) at three characteristic points very near the three minima of $W(r)$ found for this molecule (see Fig. 10). Intuitively, each point was associated with one of the lone pairs (of N or O) present in the molecule. The first column of Table, 1 which refers to a point near the N atom, indicates that the contribution of the N lone pair localized group (two electrons and two unit positive charges on N) is by far the largest (−258,6 kcal/mole). Its contribution turns out to be drastically modified by those of the other groups − the overall value is −79.6 kcal/mole: this example shows the importance of the whole molecular framework in establishing the actual value of the potential at a given point.

Further examination of the first column of the Table reveals that:

a) the contributions from the core groups (two ls electrons plus two nuclear charges) of the different atoms are practically negligible.

b) the contributions from the two oxygen lone pairs are markedly different.

The latter effect is explained by the fact that the point at which the potential is calculated lies outside the plane and is therefore asymmetrically placed with respect to these two lone pairs. The large difference between the two contributions suggests a strong directionality of the two lone-pair orbitals: in fact, as will be seen later, in a multipole expansion of the potential due to the charge distribution of such groups,

145

Table 1. $W(r)$ partition for the oxaziridine (VI) molecule in three characteristic positions in the outer molecular space[1])

Group[2])	A	B	C
$l_0(1)$	29.03	103.42	-219.44[3])
$l_0(2)$	$-$ 0.97	-218.50[3])	105.37
$1s_0$	$-$ 0.01	0.08	0.08
CH_2	23.23	21.93	21.59
b_{NH}	57.10	10.61	$-$ 4.13
b_{OC}	10.12	3.76	3.87
b_{CN}	21.83	15.53	15.14
b_{ON}	38.20	11.17	9.70
$1s_N$	0.47	$-$ 0.06	$-$ 0.06
l_N	-258.62[3])	7.62	36.46
Overall value	$-$ 79.62	$-$ 44.44	$-$ 31.42

[1]) The points selected for the analysis correspond to the minima found in the electrostatic potential, viz. A near to the N atom, and B and C near to the O atom. B is in the half-space containing the N—H bond, C in the half-space containing the N lone pair.

[2]) l_X indicates the contribution to $W(r)$ of a lone pair group of the X atom, b_{XY} the contribution due to the X—Y bond, $1s_X$ the contribution of an inner shell, while in CH_2 are collected the contributions due to both C—H bonds and $1s_C$. Values are given in kcal/mole.

[3]) The contribution from the nearest group to the selected point is written in italics.

the components of the dipole moment are noticeably important. The other two columns of Table 1 refer respectively to the analysis of $W(r)$ in the neighborhood of the O lone pair in *trans* position with respect to the N—H group and of the other O lone pair in *cis* position. The main contribution is here again due to the directly involved lone pair (-218.5 and -219.4 kcal/mole, respectively); these values are markedly decreased by the contribution of the second lone pair. Incidentally, the difference in the overall value of $W(r)$ in the two positions (-44.4 versus -31.4 kcal/mole) is due neither to the main contributions of O lone pairs nor to asymmetries in the CH_2 contributions, but almost completely to the different value of the contributions of the NH bond (b_{NH}) and the nitrogen lone pair (l_N).

Table 2. The effects of the substituents on the electrostatic potential of aziridine appreciated by means of the value of $W(\mathbf{r})$ in a given position P, near the N lone pair.

Substituent group X	CH_2	O	NH *cis*	NH *trans*
Potential of the common part[1])	-154.1	-156.0	-154.8	-155.7
W_X	21.5	28.0	19.4	32.0
$W_{b_{XN}}$	21.3 } 61.5[2])	38.2 } 76.3[2])	30.6 } 63.4[2])	29.1 } 74.0[2])
$W_{b_{XC}}$	18.7	10.1	13.4	12.9
Overall value of W	-92.6	-79.7	-91.4	-81.7
Dipole components of the XC bond				
$\mu_{\perp}$	2.31	1.55	1.83	1.83
$\mu_{\parallel}$	0	0.90	0.41	0.39

Electrostatic potentials in kcal/mole, dipole moments in Debyes.

1) $W_{1s_N} + W_{l_N} + W_{b_{NH}} + W_{b_{CN}} + W_{1s_C} + W_{b_{CH_1}} + W_{b_{CH_2}}$.

2) $W_X + W_{b_{XN}} + W_{b_{XC}}$.

In order to show more clearly the interrelations between the group potentials in a family of structurally correlated molecules, we will consider the series derived from aziridine (V) by replacing a CH_2 group with either an O atom (oxaziridine, VI) or an NH group, in *cis* as well as in *trans* position with respect to the original NH group of aziridine (*cis* and *trans* diaziridine, VII and VIII, respectively). Each of these molecules is characterized by the presence of both a NH and a CH_2 group, all involving the same type of localized orbitals ($1s_N$, l_N, b_{NH}, b_{CN}, $1s_C$, b_{CH_1}, b_{CH_2}); this is why the contributions of these orbitals to the overall potential at a point with coordinates of the $W(\mathbf{r})$ minimum lying close to the N atom[h]) have been reported as a single term in the first row of

---

h) The $W(\mathbf{r})$ trend in the series of molecules considered above presents the feature of providing the minimum near to N at practically the same local coordinates. The point chosen for an analysis of $W(\mathbf{r})$ corresponds to an average among such minimum points. Actually, the choice of one location rather than another is not particularly meaningful; on changing to a different point, one would reach equivalent conclusions, provided comparisons between different molecules were made at equivalent points.

Table 2. The three following rows give the contributions from groups which differ in the different molecules.

From the data shown in Table 2 one can readily agree about the remorkable invariance (1—2%) of the contributions of the common part. As regards the contributions of the groups which differ, the following considerations can be made:

a) The overall contribution $W_X + W_{b_{XN}} + W_{b_{XC}}$ is larger when $X \equiv O$ (+76.3 kcal/mole) in accordance with the larger electroaffinity of the oxygen with respect to the other substituting groups.

b) A comparison between *cis* and *trans* diaziridine shows how sensitive $W(\mathbf{r})$ is to the orientation of the groups: the difference between the two molecules is chiefly determined by the X contribution, *i.e.* by the whole substituting group including both the N—H bond and the lone pair, whereas the other contributions stay nearly unchanged.

c) As to the X—N bond contributions, taking the unpolarized N—N bond as reference (its potential contribution is about 30 kcal/mole), note that the $\overset{\ominus}{O}-\overset{\oplus}{N}$ bond provides a higher contribution (38 kcal/mole), and $\overset{\oplus}{C}-\overset{\ominus}{N}$ a lower one (21 kcal/mole). Intuitive considerations about bond polarity with respect to the electronegativity of the involved atoms parallel this analysis of the potential.

d) The variations in the contributions of the XC bonds, at the particular point considered, are related to the bending of these bonds rather than to their polarities, as will be shown below.

The remarks under c) and d) above may be made more specific if one examines also the sole dipole term of a multipole expansion of the bond charge distribution (see Section VIII for more details). We recall that the potential arising from a bond dipole lying in the ring plane and having components perpendicular ($\mu_\perp$) and parallel ($\mu_\parallel$) to the bond itself[1], is given by

$$V_{\text{dip}} = \mu_\perp \frac{\operatorname{sen}\vartheta \cos\varphi}{R^2} + \mu_\parallel \frac{\cos\vartheta}{R^2}$$

where R, ϑ and φ are the polar coordinates centered at the bond midpoint and having a polar axis coincident with the bond.

---

[1] For a more detailed definition, see Ref.[9]. The partition of nuclear and electronic charges adopted in that paper is the same as that employed here.

Let us consider first the X—C bonds. All the points selected for this analysis have ϑ very near to $\pi/2$, so the contribution to $W(\mathbf{r})$ given by the X—C bond is clearly due to the perpendicular component $\mu_\perp$ (which measures at the same time the bending of the bond) and not to the parallel one, $\mu_\parallel$ (which measures the polarity of the bond). The few data reported in Table 2 show, in fact, a good linearity between $W(X\text{—}C)$ and $\mu_\perp(X\text{—}C)$. Applying the same arguments to the X—N bond, we note that in this case ϑ is not far from zero ($\vartheta \simeq 12°$) and consequently the situation is reversed: the leading contribution is now given by the parallel component — $i.e.$ the charge transfer along the bond.

C. Conservation Degree of Group Potentials

The examples reported above should suffice to demonstrate the kind of considerations one may infer from breaking down electrostatic potential into group contributions. We pass now to the second point of our analysis program (see p. 143), namely to estimate the degree of conservation of group potentials. This study will be carried out in two parallel ways. In the first we shall adopt the procedure used above, $i.e.$ examination of the group considered among a given set of molecules. The second way involves verifying the degree of conservation by examining multipolar expansions of the group potential, the convergence and reliability of which will be discussed in Section VIII. For this analysis the three-membered ring molecules are again used. The groups whose conservation degree we shall consider are: CH_2 (two C—H bonds and a carbon inner shell) in the molecules III, IV, V, VI, VII, VIII, IX, X; NH (a N—H bond, a N lone pair, and a N inner shell) in the molecules V, VI, VII, VIII; and, finally, the bond C—C in the molecules III, IV, V.

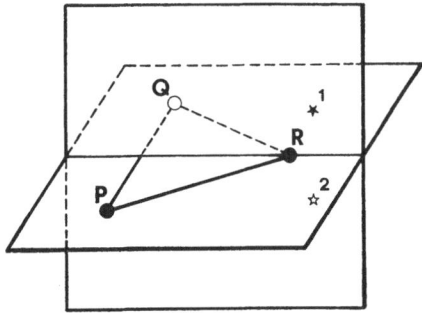

Fig. 46. Sketch of the position of the points in three-atom ring molecules selected to verify conservation properties of group potentials

The points where a check of the degree of conservation of the group potential is made have been selected to correspond to some significant positions in a group near to the one considered. In Fig. 46 the locations of the selected points 1 and 2 are marked by a star; they are adjacent to the group R — above and below the ring plane — and at a distance of about 2.3 Å from position P, where the group (CH_2 of NH) under analysis is located. C—C bonds are to be considered as located at P—Q while the location of points 1 and 2 remains unaltered.

1. CH_2 Group

The values of the electrostatic potential of the CH_2 group in position 1 give an arithmetic mean $W_{CH_2}(1) = 21.66$ kcal/mole and maximum deviation from the mean value $|\Delta W_{CH_2}(1)|_{max} = 1.15$ kcal/mole. In position 2 they are analogous: $W_{CH_2}(2) = 22.08$ kcal/mole and $|\Delta W_{CH_2}(2)|_{max} = 1.57$ kcal/mole.

The CH_2 group is not particularly sensitive to the effects of groups which decrease the symmetry; this statement is confirmed by the data in Table 3. The first two rows of Table 3 give the values of W_{CH_2} in positions 1 and 2, while the following ones contain the coefficients (in a.u.) of a multipolar expansion of $V_{CH_2}(r)$ centered at the nuclear charge center of the CH_2 group (the cartesian axes are defined at the top of Table 3). In fact, the potential $V(r)$ may be expanded as follows:

$$V(r) = \sum_a \frac{<\sigma_a>}{|r|^3} q_a + \sum_a \sum_\beta \frac{<\sigma_a \sigma_\beta>}{|r|^5} q_a q_\beta + \sum_a \sum_\beta \sum_\gamma \ldots \qquad (17)$$

In this expansion $q_a \equiv x, y, z$ represents the coordinates of the point r, where the potential is calculated, and

$$<\sigma_a> = \int \gamma(r_1, R) \sigma_a dr_1 \qquad (18)$$

$$<\sigma_a \sigma_\beta> = \int \gamma(r_1, R) \sigma_a \sigma_\beta dr_1 \qquad (19)$$

are respectively the α-th and the $\alpha\beta$-th components of the first and second moment of the molecular charge distribution. In (18) and (19), σ_a is one of the cartesian components of the vector r_1 defining the position of molecular charges.

In Table 3 we report, for brevity, only the dipole and quadrupole coefficients, although the octopole and hexadecapole terms were also considered in our analysis. The degree of constancy of such coefficients is to be considered as a measure of the shape conservation of the charge

Table 3. Electrostatic potential of the CH_2 group[1] in some molecules and the first coefficients of its multipole expansions[2]

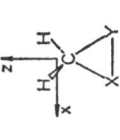

	C_3H_6	C_2H_4O	C_2H_4NH	CH_2ONH[3]	$(NH)_2CH_2$ cis	$(NH)_2CH_2$ trans	C_3H_4	N_2CH_2
$W(1)$	21.2	22.2	22.0	21.1 (O) / 21.7 (N)	21.1	22.3	22.8	20.5
$W(2)$	21.2	22.2	23.2	21.4 (O) / 22.6 (N)	22.4	22.3	22.8	20.5
$\langle x \rangle$	0	−0.021	−0.070	−0.038	0	0	0	0
$\langle y \rangle$	0	0	−0.017	−0.021	−0.037	0	0	0
$\langle z \rangle$	−0.802	−0.795	−0.796	−0.776	−0.792	−0.798	−0.812	−0.719
$\langle xx \rangle$	−2.977	−2.995	−2.987	−3.020	−3.007	−3.005	−2.964	−3.028
$\langle xy \rangle$	0	0	−0.009	0.011	0	0.021	0	0
$\langle xz \rangle$	0	−0.006	−0.018	0.010	0	0	0	0
$\langle yy \rangle$	−3.188	−3.237	−3.246	−3.268	−3.284	−3.296	−3.267	−3.189
$\langle yz \rangle$	0	0	−0.020	−0.016	−0.035	0	0	0
$\langle zz \rangle$	−2.771	−2.630	−2.661	−2.607	−2.633	−2.634	−2.642	−2.634

1) Calculated at the two points 1—2 defined in Fig. 46; the values are in kcal/mole.
2) Values in a.u.
3) In the oxaziridine molecule, two different definitions of points 1—2 are possible: near the O atom or the NH group. Both are reported in the table.

distribution and consequently of the corresponding electrostatic potential. With reference to the present choice of axes, the coefficients $<x>$, $<y>$, $<xy>$, $<xz>$, $<yz>$ gauge the deformation with respect to the symmetry C_{2v} of V_{CH_2} induced by the neightboring groups. The extent of variation of the other terms among the set of molecules is a clue to the degree of conservation. An examination of such data shows a fair conservation of the electrostatic potential of CH_2; analogous invariance has also been found for other properties of this group in the same set of molecules[9,12].

It may be as well to emphasize here that the values in Table 3 refer to minimal basis set wave functions, and that the conclusions we have drawn are to be considered as provisional until equivalent analyses in terms of more flexible wave functions become available. To the extent that these results are satisfactory, we can define, a set of mean values for the expansion coefficients, since the spread of the values is sufficiently limited. — These mean values, up to the octopole terms, are reported in Table 4 and can be used in Eq. (17) for approximate calculations of the electrostatic potential.

Table 4. Mean values of the expansion coefficients for the CH_2, NH, and C—C groups in three-atom rings

	CH_2	NH	C—C
$<x>$	0	0	0.886
$<y>$	0	0.725	0
$<z>$	−0.786	−0.916	0
$<xx>$	−2.998	−2.258	−1.857
$<xy>$	0	0	0
$<xz>$	0	0	0
$<yy>$	−3.247	−2.523	−1.383
$<yz>$	0	0.548	0
$<zz>$	−2.652	−2.753	1.273
$<xxx>$	0	0	2.149
$<xxy>$	0	−0.655	0
$<xxz>$	−1.083	−0.637	0
$<yyx>$	0	0	0.678
$<yyy>$	0	−0.096	0
$<yyz>$	0.307	−0.135	0
$<zzx>$	0	0	−0.112
$<zzy>$	0	−0.323	0
$<zzz>$	−2.403	−1.817	0
$<xyz>$	0	0	0

The coefficients are given in a. u.

2. NH and C—C Groups

There is little point in an analogous discussion of the other two groups, NH and b_{C-C}. The analysis of the variation of the potential at the two selected points 1,2 among the set of molecules gives the following results for the NH group:

$$W_{NH}(1) = 30.94 \text{ kcal/mole} \qquad |\Delta W_{NH}(1)|_{max} = 1.18 \text{ kcal/mole};$$
$$W_{NH}(2) = 18.10 \text{ kcal/mole}, \qquad |\Delta W_{NH}(2)|_{max} = 1.26 \text{ kcal/mole}.$$

The few values available for the C—C bonds clearly indicate that this potential contribution is strictly related to the strain of the bond, as stated above for the X—C bonds. The values we have for W_{C-C} are the following: 14.74 kcal/mole for cyclopropane, 20.43 kcal/mole for oxirane (in both molecules positions 1 and 2 are equivalent), 18.96 kcal/mole and 18.74 kcal/mole for positions 1 and 2 of aziridine. They run parallel to the perpendicular component $(\mu_\perp)$ of the C—C bond dipole moment; the corresponding values are: $\mu_\perp = 2.02$ D for cyclopropane, 2.42 D for oxirane and 2.31 D for aziridine. One may conjecture that in compounds where the strain of the bond does not play such an essential role the degree of conservation of W_{C-C} will be better.

As far as the multipolar expansion (17) is concerned, both groups have coefficients which change on passing from one molecule to another to the same extent as those of the CH_2 group. The mean values of these coefficients are reported in Table 4.

VIII. Analytical Approximations of $V(r)$

A. One- and Many-center Multipole Expansions

An analytical expansion of the electrostatic molecular potential could be very useful for actual utilizations of the electrostatic approximation. The choice of the best analytical form is in general dictated by the specific problem. For the applications considered in the present paper we have selected multipole expansions into spherical harmonics:

a) referred to a unique center (the center of the nuclear charges of the molecule)

b) referred to as many centers as there are localized orbitals. Each expansions concerns a single LO and the expansion center coincides with the center of nuclear charges associated with that LO.

c) referred to as many centers as are the chemical groups we consider the molecule to be partitioned.

153

The first problem we have to deal with is related to the convergence properties of the chosen analytical expression. We are not interested, however, in the formal aspects of this question and limit ourselves to the practical implications of some actual results.

Let us consider first a small molecule, ammonia, for which the most obvious expansion choice is the one-center one. To verify the behaviour of the a) expansion with the distance, a set of concentric spheres (centered at the nuclear charge center) was selected, and on each sphere a set of 614 equally spaced points [50]. The electrostatic potential was calculated for each of these points, according to either the exact SCF formula (Eq. 13) or the multipole expansion (as in Eq. 17). The expansion was progressively extended, first including only the dipole term, then the quadrupole, the octopole and finally the hexadecapole ones. The mean values of $W(\mathbf{r})$ on each sphere are reported in Fig. 47 as a function of the radius R. For distances greater than 2 Å, an expansion truncated after the octopole terms may be considered sufficiently accurate.

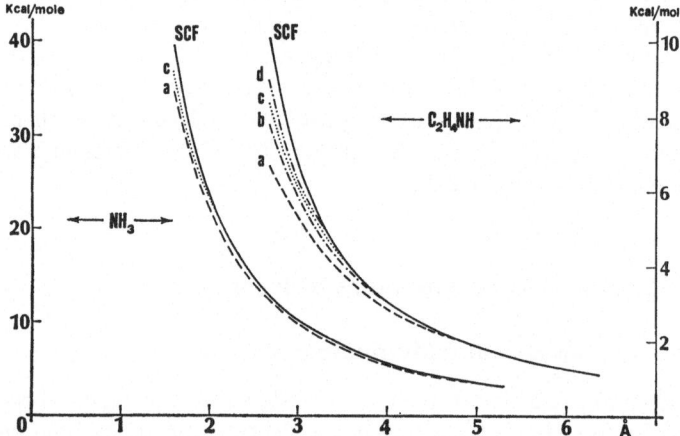

Fig. 47. Convergency properties of the multipole expansion of electrostatic energy. $W(R)$ is the mean value of electrostatic potential on a sphere of radius R calculated by a) dipole expansion terms only, b) dipole and quadrupole, c) dipole, quadrupole and octopole terms, d) as c) plus hexadecapole terms; e) SCF values. The left side of figure refers to ammonia, the right one to aziridine

In a larger molecule convergence will be slower. On the right side of Fig. 47 similar results are reported for the aziridine molecule. The expansion containing terms up to octopole is sufficiently accurate for distances more than 3 Å from the center of the molecule. Another test is to analyze the errors introduced by using the multipolar expansion.

The percentage mean value of the absolute deviations between exact (SCF) and approximate (expansion up to hexadecapole terms) electrostatic potentials on the $N = 614$ points was calculated for each of the spheres:

$$\eta = \frac{100}{N} \sum_{i=1}^{N} \frac{\left| V^{\text{SCF}}(\boldsymbol{r}_i) - V^{\text{app}}(\boldsymbol{r}_i) \right|}{\left| V^{\text{SCF}}(\boldsymbol{r}_i) \right|} \tag{20}$$

The curves refer to H_2O (a), NH_3 (b), aziridine (c) (solid curves). For ease of visualization, η is reported on a logarithmic scale, while distances are measured in fractions of the medium van der Waals molecular radius of the molecule considered. The η values rapidly decrease with increasing distance and, in the case of the two small molecules. they are less than 10% at a distance equal to $\overline{R}_{\text{vdW}}$. With aziridine, it is necessary to go to distances larger than 1.4 $\overline{R}_{\text{vdW}}$ in order to have comparable η values. Fig. 48 gives analogous results obtained from a many-center expansion of the b) type (dashed curves). As was to be expected, at

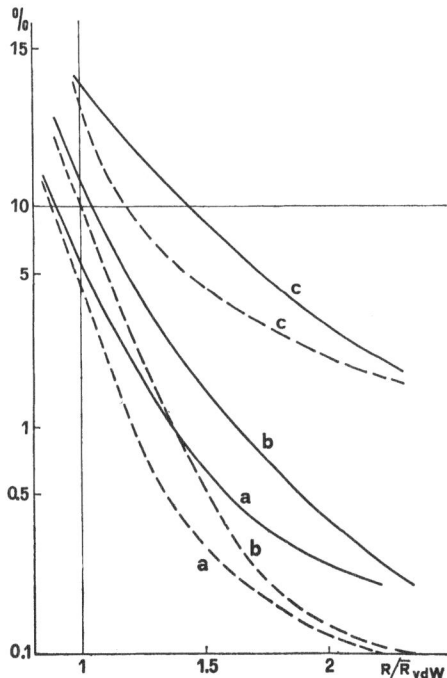

Fig. 48. Variation with the distance of the percentage mean value of the error for one-center (solid lines) and many-center (broken lines) expansions of the electrostatic potential: a) water, b) ammonia, c) aziridine. From Ref. [50]

medium distances the many-center approximation gives better agreement.

Some results related to the c) type expansions have already been reported in Section VII. The convergence of such expansion is, of course, intermediate between a) and b).

A more complete analysis of the convergence properties of multipole expansions would require an examination of their performance in each orientation. To report the pertinent data would take too long, so we refer the reader different check of convergence discussed in Section IX. B.

Recently Pack, Wang and Rein[51] published a convergence analysis of analytical expansions of the electrostatic potential on parallel lines to the present one. These authors compare with the exact expansion the one-center one and a "segmental atomic" expansion (centered at the nuclei). Convergence is tested on pyridine (semiempirical iterative extended Hückel wave function) along the symmetry axis with expansion truncated after the octopole term. Their results are comparable to those reported here; in particualr, the segmental expansion appears quite reasonable[J].

One-center expansions of the electrostatic potential have been employed in theoretical studies on electron scattering by diatoms. A single-center basis set was used by Ardill and Davidson[53] for H_2; two-center basis sets with subsequent expansion of the MO's into single-center molecular orbitals were used by Faisal[54] for N_2, and by Gianturco and Tait[55] for CO — in the last paper the results of two different wave functions are compared; numerical calculation of expansion coefficients was performed by Thruhlar, Van-Catledge and Dunning[56] for different wave functions of N_2, ranging from the HF limit to semiempirical INDO ones.

B. Monopole Expansions

Another way to obtain an analytical expression for $V(r)$ is by a point charge representation of the molecular charge distribution. Such a procedure can be of practical use only if the number of point charges is reasonably limited. In our experience, it is quite difficult to get a sufficiently accurate representation of the charge distribution for medium-sized molecules,, whereas for small molecules, like water and ammonia, it is relatively easy to do so.

[J] Riera and Meath[52] agree in considering generally insufficient expansions truncated after the quadrupole terms.

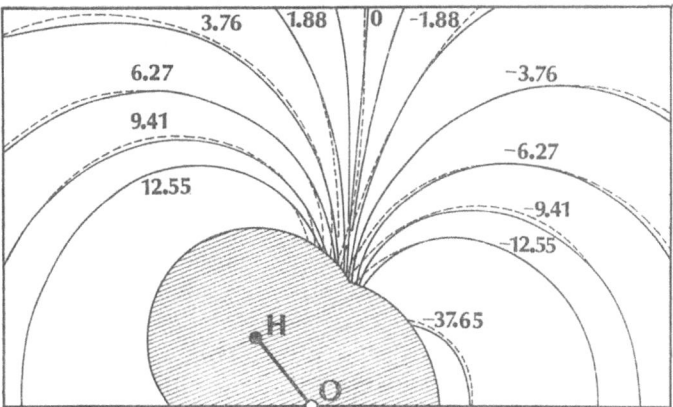

Fig. 49. Comparison of SCF and point-charge model values of $W(\mathbf{r})$ in the molecular plane for H_2O. Solid line refers to SCF values, dashed lines to the model. From Ref. [57]

Fig. 49 shows by way of example a comparison between the SCF potential and that obtained by 13 point charges[57] for the H_2O molecule. The geometrical arrangement of the charges was chosen so as to minimize the deviations with respect to the SCF potential at a fairly large number of points of the outer molecular space, with the simultaneous constraint that the point charge distribution must give the same value of dipole and quadrupole components as the SCF wave function. For more information about the locations of the points, reference should be made to the original paper[57]. For the regions outside the molecular van der Waals surface, the monopole expansion gives values of $W(\mathbf{r})$ which are accurate enough for chemical applications in the molecular plane (as shown in Fig. 49) as well as in the other portions of the outer molecular space (not reported here).

IX. Electrostatic Description of the Conformational Structures of Molecular Associates $A \cdot H_2O$

A. Direct Application of the SCF Electrostatic Potential

The monohydration associates $A \cdot H_2O$, where A is a neutral molecule containing polar groups, represent a typical association to which the electrostatic approximations of Section II. E may be applied. The object is to obtain by relatively inexpensive methods a first-order de-

157

scription of the interaction energy surface between the two partners. We would be satisfied with predictions about the location of the solvation sites of A, the approximate geometry of stable associates and their conformational, energies, and a semi-quantitative and comparative estimate of mono-solvation energies at the various possible sites.

One may infer from experimental data that this type of association — essentially a hydrogen-bond association — does not involve critical variations of the internal geometry of the partners[58]. Consequently, it should be legitimate to use in our description the wave functions of the separate molecules. On the other hand, some analyses performed on SCF calculations of hydrogen-bonded associates[59-61] show that at distances beyond the equilibrium ones the interaction energy is mainly of electrostatic nature, but at the actual equilibrium distance the differences between the total interaction energy and the electrostatic portion appear to be relevant. The application of electrostatic methods, therefore, does not encounter particularly serious problems in evaluating medium- and large-distance interactions, whereas at distances closer to the equilibrium ones, things are more complicated. Moreover, the above-mentioned analyses clearly show that the electrostatic approximation is not able to predict the equilibrium distances correctly. However, we may ask whether at fixed $R (\geqslant R_{eq})$ the electrostatic approximation can indicate the correct trend of the energy as a function of the other parameters describing the mutual orientation of the two molecules; if it could, a partial, first-order answer might be given to the other questions about the associate.

We must point out that, if the assumption of internal rigidity is accepted for both associating molecules, the number of degrees of freedom of the system $A \cdot H_2O$ shrinks to six: three parameters are required to fix the position of a point in H_2O (*e.g.* the O nucleus) with respect to the other molecule, and three further parameters are needed to specify the H_2O orientation with respect to A. Thus, the interaction energy hypersurface $W_{A \cdot H_2O}$ is defined in a six-dimension configurational space and it is obviously impossible to visualize its shape unless one resorts to examing particular (and significant) "sections" of such a hypersurface.

With a view to checking whether or not the results of the electrostatic calculations performed on $W_{A \cdot H_2O}$ section with constant R are able to provide a sufficiently reliable picture, we shall compare a few "sections" obtained by the approximate method with some relying upon the usual *ab initio* SCF calculation (in terms of a minimal basis set). In the electrostatic calculations, since one of the partners is always an H_2O molecule, the point charge description of paragraph VIII. B may be utilized. The interaction energy W will consequently be obtained by applying Eq. (9):

$$W_{A \cdot H_2O} = \sum_{k=1}^{n} V_A(\textbf{k}) \, q_{kH_2O} \qquad (21)$$

which utilizes, for each configuration of the associate, the electrostatic potential of A in the $n = 13$ points where the q_k charges of water are placed.

The most thoroughly investigated $A \cdot H_2O$ system — using *ab initio* methods — is the water dimer[57,62-68]. In this particular case, the electrostatic method has given encouraging results (for a description see the original paper[57]). Here we will consider a slightly more complex example, the adduct formamide–water; this is more interesting because formamide has two polar groups and several possible monosolvation sites. A detailed study[69] has shown that the SCF procedure and the electrostatic method agree in forecasting five preferred hydration sites, all located in the formamide plane; these are summarized in Fig. 50.

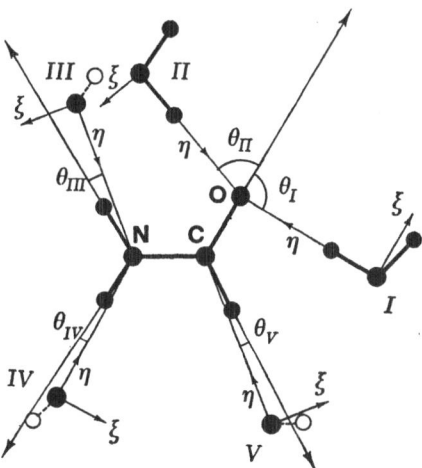

Fig. 50. A sketch of the geometries of the five most important associations $H_2NCHO \cdot H_2O$. From Ref. [69]

The location of all these sites is near, but not completely coincident with, what one could have predicted on an intuitive basis. In particular, in the associates where H_2O acts as a proton acceptor, the hydrogen bond $XH \cdots OH_2$ is not rigorously linear, and in the associates where H_2O acts as a proton donor, the location of the water molecule is not directly inferred from the known direction of the carbonyl oxygen lone pairs. Intuitive arguments, in fact, rely upon a tacit consideration of local fields caused by the group directly involved in the association, but the effects due to the whole molecular context are important (see Section VII. B). The individuation of the hydration sites is not sufficient to completely characterize the associate. Sections of the $W_{A \cdot H_2O}$ hypersurface corresponding to the three parameters defining the orientation of the water molecule (we have employed for this purpose the angles of rotation around three orthogonal local axes ξ, η, ζ, see Fig. 50) will give the essential part of the additional information we need.

Fig. 51 shows the three orthogonal sections of $W_{A \cdot H_2O}$ concerning the associate II (defined in Fig. 50). Solid curves refer to SCF calculations, dashed curves to electrostatic calculations. A satisfactory agreement between the two sets of results is evident, both as regards the shape of the curve — useful for further thermodynamic characterizations of the associate — and the location of the minima. In particular, the rotation around the ζ axis shows that SCF and electrostatic calculations agree in forecasting a non-linearity of the hydrogen bond of about 12°. The

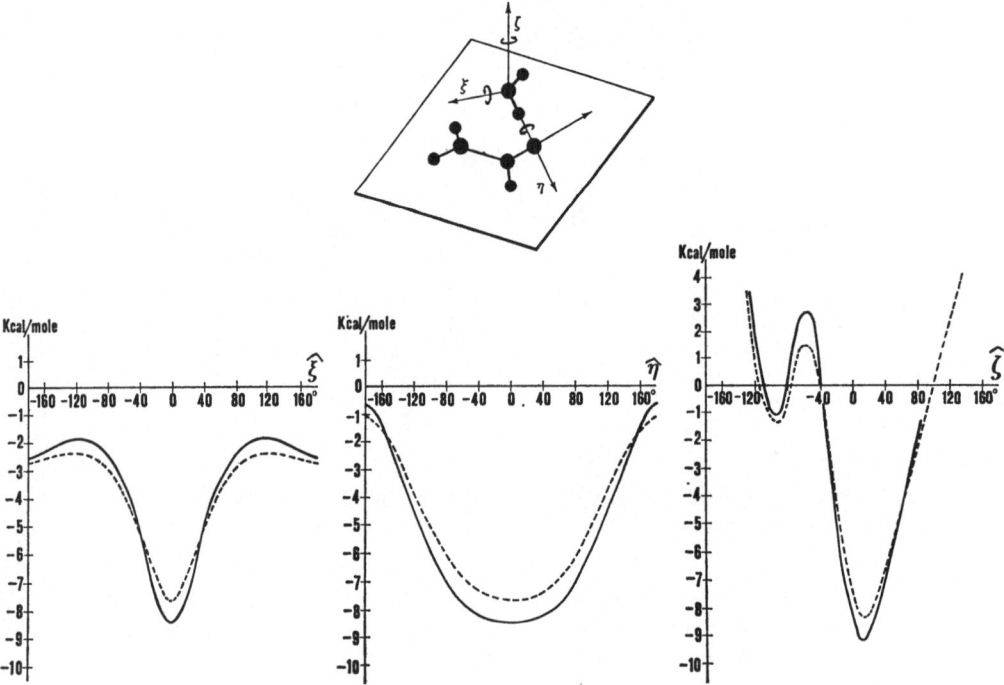

Fig. 51. Variations of the conformational energy for association II of $H_2NCHO \cdot H_2O$. a) Rotation of H_2O around the ξ axes. b) Rotation around the η axis (*i.e.* around the hydrogen bond). c) Rotation around the ζ axis (non-linearity of the hydrogen bond). Solid line: SCF results; dashed line: model. From Ref. [69]

two sections reported in Fig. 57 (the third is not of particular interest) refer to associate III, where H_2O acts as a proton acceptor. The shape of the SCF representation of the $W_{A \cdot H_2O}$ surface is different from that of associate II; these differences are reasonably well accounted for by electrostatic calculations.

The best values of the conformational parameters of the five most stable associates are collected in Table 5. By examing this table, the reader may judge the degree of confidence he may assign to such electrostatic predictions. As for us, we note that the electrostatic model gives worse results when H_2O acts as a proton donor as is clear from a comparison of SCF and electrostatic stabilization energies: the electrostatic results for associates I and II do not fit with the others.

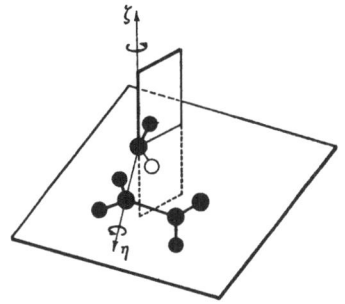

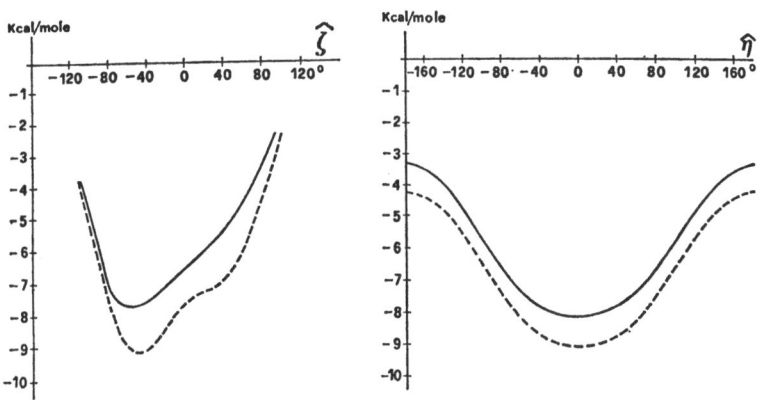

Fig. 52. Variations of the conformational energy for association III of $H_2NCHO \cdot H_2O$. a) Rotation of H_2O around the η axis. b) Rotation around the ζ axis. Solid line: SCF results; dashed line: model

161

It is a matter for further investigation to establish whether such failures are due entirely to an intrinsic inadequacy of the electrostatic assumption or depend to some extent on the further approximation we introduced, *i.e.* representing the charge distribution of water by means of a limited set of point charges.

B. Application of the Analytical Expansions of $V(r)$

The electrostatic method, as we have seen, shows some inconsistencies when association energies of different types of monohydration associates are compared, but still gives a sufficiently good description of the confor-

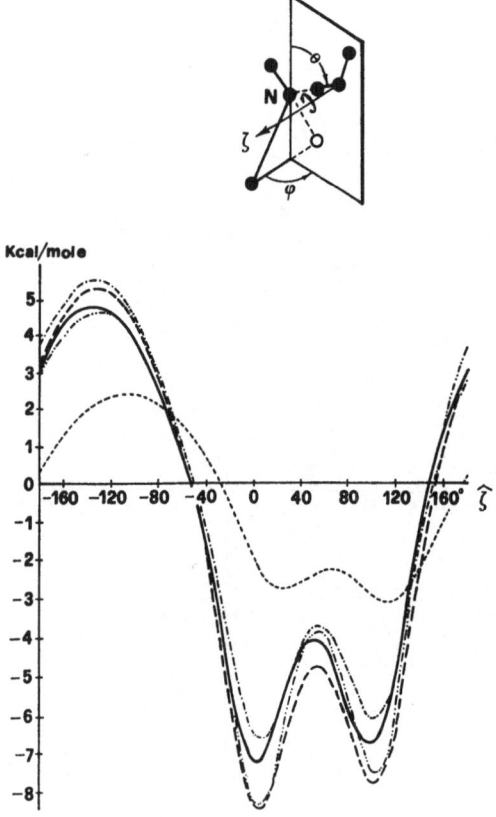

Fig. 53. Comparison between "exact" electrostatic and approximate descriptions of conformational energy for an aziridine-water associate. Solid, line: "exact" electrostatic potential; --- expansion truncated after dipole terms; - · - · - after quadrupole terms; - · · · · · after octopole terms; --- after hexadecapole terms. From Ref. [50]

Table 5. Geometries, barrier heights and stabilities for the most interesting water-formamide mono-associates (from Ref. [69])

Associate[1]	Configurational parameters			Conformational parameters			Barrier[3] heights	Stabilization energies
	$R^{[2]}$ (Å)	ϑ (deg)	φ (deg)	$\hat{\zeta}$ (deg)	η (deg)	$\hat{\zeta}$ (deg)	(kcal/mole)	(kcal/mole)
I	2.815	88	180	0	0	− 7	5.0	9.4
	(2.815)	60	180	0	0	−10	3.4	6.5
II	2.815	72	0	0	180	12	7.8	9.2
	(2.815)	67	0	0	180	13	6.6	8.3
III	2.83	10	180	0	0	−55	4.8	8.2
	(2.83)	10	180	0	0	−47	4.9	9.4
IV	2.85	5	0	0	0	−18	0.4	7.2
	(2.85)	5	0	0	0	−32	0.4	8.4
V	3.25	6	180	0	0	72	2.1	3.0
	(3.25)	4	180	0	0	60	2.1	3.5

[1] See Fig. 50 for the definition of the associates. For each associate, the numbers in standard characters are the SCF results, the numbers in italics correspond to the electrostatic model.

[2] For the model. the values given refer to computations made at equilibrium SCF distances.

[3] Barrier heights for rotations around the hydrogen-bond axis.

163

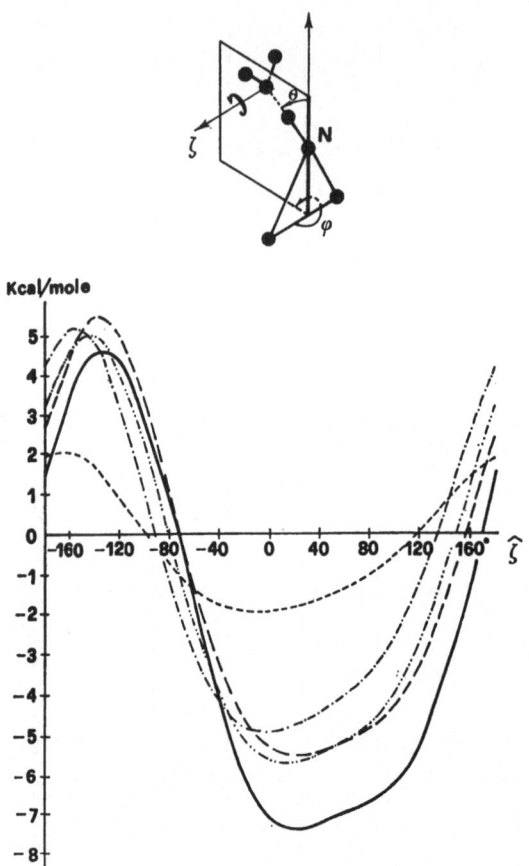

Fig. 54. As in Fig. 53 for a different aziridine-water associate. From Ref. [50]

mation energy. The question is, is it possible to resort to the analytical expansions of $V(r)$ introduced in Section VIII in order to reduce computer time for calculating conformational energy surfaces of associates. In the Section VIII. C it was noted that, for molecules of intermediate complexity, the analytical expansions of $V(r)$ we have examined begin to be satisfactory at distances greater than about 1.5 times the van der Waals molecular radius. In consequence, a preliminary investigation is to be made of the reliability of conformational calculations performed on a somewhat shorter distance.

Fig. 53 depicts a section of the conformational energy surface for an aziridine–water associate: in this case the water acts as a proton acceptor (see sketch at top of figure). The results of a calculation performed directly by means of Eq. (21) (solid-line[70]) are compared with those obtained by using the one-center expansion of $V(\mathbf{r})$ of aziridine: to give an idea of the convergence rate of such an expansion, expansions truncated after the dipole, quadrupole, octopole and hexadecapole terms are reported.

In the example given in Fig. 54 the water molecule acts as a proton donor in the $C_2H_4NH \cdot H_2O$ associate. The comparisons show that the octopole curves are sufficient to reproduce the trend of the "exact" electrostatic curves, although they are rather far from giving the same numerical value for the interaction SCF energy. In conclusion, we may say that analytical expansions seem adequate to reveal the main characteristics of the conformational surface; verifications are, of course, necessary.

Note added in proofs. It should be clear that this paper was not designed to offer a review of the applications of the electrostatic approximations to the chemical reactivity or molecular interaction problems. However, it may be of some interest to add a few quotations of further developments and direct applications of the electrostatic molecular potential method performed or noticed by the authors after the completion of the present paper.

A. B. Anderson[71] has recently shown that generally the interaction between a molecule and a charged atom (both taken in a frozen form) can be exactly obtained within the limits of the Hellmann-Feynman approximation, giving the same result of Eq. (13).

Independently, Srebrenik, Weinstein and Pauncz[72] remarked the relation between the definition of the molecular potential $V(\mathbf{r})$ and the Hellmann-Feynman theorem. Consequently, by analytic integration of the related Poisson equation:

$$\nabla^2 V(\mathbf{r}) = -4\pi\gamma(\mathbf{r})$$

these authors have been able to give a method particularly convenient for computing $V(\mathbf{r})$ when the basis set is given by GTO's (for STO's an expansion over gaussian functions is always possible).

Several attempts to extend to shorter distances between reactants the utilization of such simple methods, are at present under examination. Such attempts in some manner approximate other portions of the interaction energy: charge transfer, exchange, polarization (see Chap. II). In particular, F. P. Van Duijneveldt[73] estimates the polarization energy due to an approaching proton according to the Rayleigh-Schrödinger perturbation scheme, truncated at the second order:

$$E_{pol}(A-H^+) \simeq -2 \sum_{i}^{occ} \sum_{k}^{vir} \frac{|<\phi_{iA}|r_{\bar{H}}^{-1}|\phi_{kA}>|^2}{\varepsilon_i - \varepsilon_k}$$

using the same one-electron integrals already used to calculate $W(\mathbf{r})$ (Eq. 13). For the ethylene protonation process, such corrections permit a neater definition of the approaching path, with a better agreement with full *ab initio* calculations.

Compounds of noticeable chemical interest have been studied at the Stockholm University in the SCF *ab initio* framework. In the thiophene molecule [74], reactive areas for electrophilic reactants have been found in the π region above and below the C—C double bonds and the sulphur atom (the electrons around S have a tetrahedral rather than a trigonal arrangement). The authors consider as more likely an electrophilic attack proceeding *via* weak intermediate addition complexes of π-type and estimate the ratio of C_α/C_β substitution in a good agreement with experimental data.

In another paper [75] the same group considers benzene and aza-benzenes (pyridine, pyridazine, pyrimidine, pyrazine, s-triazine, s-tetrazine). The results for pyridine and pyrazine parallel those reported in the present paper. The examination of the overall set of results suggests some hints concerning reactivity and reaction mechanisms. For the electrophilic substitutions at a carbon atom, the benzene molecule appears the most reactive (a bridged carbonium ion as an intermediate in the attack may be suggested). The nucleoplic character of the carbon atoms drastically decreases when one or more nitrogen atoms are introduced in the ring, in accordance with experiment. For the protonation of a nitrogen atom the authors evidence a good correlation of the potential minima with the relative basicities for the diazines, while the electrostatic potential does not explain why the pK_a of pyridine is far greater than that of diazines. Perhaps such a fact may be due to the reasons we have outlined at p. 131.

In a third paper [76], the same group of researchers considers the electrophilic substitutions in fluorobenzene: the outstanding feature of such a compound, *i.e.* to have a para-directing effect, is accounted for by the electrostatic molecular potential.

In the fields of molecular associations, G. Port and A. Pullman have determined the location of the main hydration sites in the purinic and pyrimidinic bases of nucleic acids [77]. An expansion of the electrostatic potential somewhat different from those reported in Chap. VIII was employed [78]. The results show that association with a water molecule is preferred in every case on the ring plane, with well evidenced minima.

In these last months some applications of the electrostatic molecular potential approach have been performed. Such studies concern β-adrenergic drugs and psycotomimetic cholinergic agents. In the first family, compounds like isoproterenol and INPEA have been examined in the CNDO approximation [79,80]. In the second family a large set of compounds have been considered in the INDO approximation. An incomplete list includes: pseudotropine, scopolamine, atropine, acetoxytropine, acetoxypseudotropine, 3-acetoxyquinuclidine, N-methyl-3-acetoxy-quinuclidinium, acetoxy-cyclopropyltrimethylammonium, trihexyphe-nidyl, a set of 1-cyclohexylpiperidine derivatives (1-phenyl, 1-ethynyl, 1-acetonitrile) [81,82,83]. It is surely untimely to draw any conclusion, but it is likely to foresee that the number of applications of this electro-static method to molecular pharmacology will increase in the next future.

An extension of the approach here outlined to excited state species is at present in progress. As a first example, the potential maps of the first singlet and triplet excited states of thymine using the C. I. wave functions of Snyder, Shulman and Neuman [84] have been published [85]. Large changes in the reactivity of the chemical groups of the molecule are evident. Moreover, the partition of the interaction energy for molec-ular associations involving excited species, performed recently by K. Morokuma [86] seems to show that the electrostatic portion accounts reasonably well for the conformational energy. Of consequence, the electrostatic procedures elaborated for associations in the ground state seem to be potentially useful also for predicting the orientation of the partners in excited state associations.

Acknowlegments. The authors are particularly grateful to Dr. R. Bonaccorsi for her help in performing the calculations This work was supported by the Laboratorio di Chimica Quantistica ed Energetica Molecolare del CNR.

X. References

1) Boys, S. F., Rajagopal, P.: In: Advances in quantum chemistry, Vol. 2 (ed. P. O. Löwdin) p. 1. New York: Academic Press 1965.

2) Parr, R. G.: Quantum theory of molecular electronic structure. New York: Benjamin 1964. — Mc Weeny, R., Sutcliffe, B. T.: Methods of molecular quantum mechanics. New York: Academic Press 1969.

3) For recent reviews, see: Hirschfelder, J. O. (ed.): Advances in Chemical Physics *12*. New York: Wiley 1967. — Margenau, H., Kestner, N. R.: Theory of Inter-molecular Forces. New York: Pergamon Press 1971. — Certain, P. R., Bruch, L. W.: In: MTP Intern. Rev. of Sci., Vol. 1, Theoretical Chemistry (ed. W. Byers Brown), p. 113. London: Butterworths 1972.

4) Examples of computational procedures having such characteristic are: Magnasco, V., Dellepiane, G.: Ric. Sci., Parte 2, Sez. A *33*, 1173 (1963); Ric. Sci. Parte 2, Sez. A. *34*, 275 (1964). — Musso, G. F., Magnasco, V.: J. Phys. B., Atom. Mol. Phys. *4*, 1415 (1971); 1s functions only. — Guidotti, C., Maestro, M., Salvetti, O.: Ric. Sci., Parte 2, Sez. A *35*, 1155 (1965). — Guidotti, C., Salvetti, O., Zand-omeneghi, M.: Ric. Sci., Parte 2, Sez. A *36*, 25 (1966). — Guidotti, C., Maestro, M., Salvetti, O.: Ric. Sci., Parte 2, Sez. A *37*, 234. — Guidotti, C., Salvetti, O.,

Zandomeneghi, M.: Ric. Sci., Parte 2, Sez. A *37*, 240 (1967); functions with $n \leqslant 6$, $l \leqslant 2$. — McLean, A. D., Yoshimine, M.: IBM J. Res. Develop. 12, 206 (1968); linear molecules only. — Wahl, A. C., Land, R. H.: Intern. J. Quantum Chem. *1 S* 375 (1967) — Wahl, A. C., Land, R. H.: J. Chem. Phys. *50*, 4725 (1969); functions with $n \leqslant 8$, $l \leqslant 7$.

[5] Bonaccorsi, R., Scrocco, E., Tomasi, J.: Theor. Sect. Prog. Report pag. 35. Pisa, Lab. Chim. Quant. CNR 1970; see also: Aspects de la Chimie quantique contemporaine, ed. by R. Daudel and A. Pullman, p. 82. Paris, C. N. R. S. 1971.

[6] Frichtie, C. J.: Acta Cryst. *20*, 27 (1966). — Hartmann, A., Hirshfeld, F. L.: Acta Cryst. *20*, 80 (1966).

[7] Coulson, C. A., Moffitt, W. E.: Phil. Mag. *40*, 1 (1949). — Coulson, C. A., Goodwin, T. H.: J. Chem. Soc. *1962*, 2851. — Peters, D.: Tetrahedron *19*, 1539 (1963). — Veillard, A., Del Re, G.: Theoret. Chim. Acta *2*, 55 (1964). — Klasinc, L., Maksić, Z., Randić, M.: J. Chem. Soc. *A 1966* 755. — Bernett, W. A.: J. Chem. Educ. *44*, 17 (1967).

[8] Walsh, A. D.: Nature *159* 165, 712 (1947). — Walsh, A. D.: Trans. Faraday Soc. *45*, 179 (1949). — Sugden, T. M.: Nature *160*, 367 (1947).

[9] Bonaccorsi, R., Scrocco, E., Tomasi, J.: J. Chem. Phys. *52*, 5270 (1970).

[10] Petke, J. D., Whitten, J. L.: J. Am. Chem. Soc. *90*, 3338 (1968).

[11] Radom, L., Pople, J. A., Buss, V., Schleyer, P. v. R.: J. Am. Chem. Soc. *94*, 311 (1972).

[12] Bonaccorsi, R., Scrocco, E., Tomasi, J.: Theoret. Chim. Acta *21*, 17 (1971).

[13] For some recent reviews, see: Vol'pin, M. E.: V. Int. Conf. on Organomet. Chem., Moscow 1971: Plenary Lectures. London: Butterworths 1972. — Chatt. J., Leigh, G. J.: Chem. Soc. Rev. *1*, 121 (1972).

[14] Bonaccorsi, R., Scrocco, E., Tomasi, J.: unpublished results.

[15] Bonaccorsi, R., Pullman, A., Scrocco, E., Tomasi, J.: Chem. Phys. Letters *12*, 622 (1972).

[16] Alagona, G., Pullman, A., Scrocco, E., Tomasi, J.: unpublished results.

[17] Pullman, A.: Chem. Phys. Letters *20*, 29 (1973). — Hopkinson, A. C., Csizmadia, I. G.: J. Can. Chem. *51*, 1432 (1973).

[18] Homer, R. B., Johnson, C. D.: In: Chemistry of amides (ed. J. Zabicky). New York: Wiley 1970.

[19] Kirby, A. H. M., Neuberger, A.: Biochem. J. *32*, 1146 (1938).

[20] Dedichen, G.: Ber. *39*, 1831 (1906).

[21] Brown, D. G., Gosh, P. B.: J. Chem. Soc. *1969 B*, 270.

[22] Berthier, G., Bonaccorsi, R., Scrocco, E., Tomasi, J.: Theoret. Chim. Acta *26*, 101 (1972).

[23] See, *e.g.*: Kochetkov, N. K., Sokolov, S. D.: Advan. Heter. Chem. *2*, 365 (1963). — Kost, A. N., Grandberg, I. I.: Advan. Heter. Chem. *6*, 347 (1966). — Grimmett, M. R.: Advan. Heter. Chem. *12*, 104 (1970). — Albert, A.: Heterocyclic chemistry. London: Athlone Press 1968.

[24] Chiang, Y., Whipple, E. B.: J. Am. Chem. Soc. *85*, 2763 (1963).

[25] Albert, A., Goldacre, R., Phillips, J. N.: J. Chem. Soc. *1948*, 2240.

[26] Chia, A. S., Trimble, R. F.: J. Phys. Chem. *65*, 863 (1961).

[27] Schofield, K.: Hetero-aromatic nitrogen compounds, p. 270. London: Butterworths 1967.

[28] Cheeseman, G. W. H., Werstink, E. S. G.: Advan. Heter. Chem. *14*, 99 (1972).

[29] Bonaccorsi, R., Pullman, A., Scrocco, E., Tomasi, J.: Theoret. Chim. Acta *24*, 51 (1972).

[30] Christensen, J. J., Rytting, J. H., Izatt, R. M.: Biochemistry *9*, 4907 (1970).

[31] Pal, B. C.: Biochemistry *1*, 558 (1962).

32) Clementi, E., Raimondi, D. L.: J. Chem. Phys. *38*, 2686 (1963).
33) Clementi, E., Clementi, H., Davis, D. R.: J. Chem. Phys. *46*, 4725 (1967).
34) Mély, B., Pullman, A.: Theoret. Chim. Acta *13*, 278 (1969).
35) Arrighini, G. P., Guidotti, C., Salvetti, O.: J. Chem. Phys. *52*, 1037 (1970).
36) Giessner-Prette, C., Pullman, A.: Theoret. Chim. Acta *25*, 83 (1972).
37) Aung, S., Pitzer, R. M., Chan, S. I.: J. Chem. Phys. *49*, 2071 (1968).
38) Bonaccorsi, R., Tomasi, J.: unpublished results.
39) Ghio, C., Tomasi, J.: Theoret. Chim. Acta *30*, 151 (1973).
40) Petrongolo, C., Tomasi, J.: Chem. Phys. Letters *20*, 201 (1973).
41) Hehre, W. J., Stewart, R. F., Pople, J. A.: J. Chem. Phys. *51*, 2657 (1969).
42) De Paz, M., Leventhal, J. J., Friedman, L.: J. Chem. Phys. *51*, 3748 (1969). —
Long, J., Munson, B.: J. Chem. Phys. *53*, 1356 (1970).
43) Bonaccorsi, R., Scrocco, E.: unpublished results.
44) Mély, B., Pullman, A.: Compt. Rend. *274*, 1371 (1972).
45) Weinstein, H., Pauncz, R., Cohen, M.: In: Advances in atomic and molecular
physics (ed. D. R. Bates and I. Esterman), Vol. 7, p. 97. New York: Academic
Press 1971.
46) Foster, J. M., Boys, S. F.: Rev. Mod. Phys. *32*, 300 (1960).
47) Ruedenberg, K.: In: Modern quantum chemistry, (ed. O. Sinanoğlu), Part 1,
p. 85. New York: Academic Press 1965.
48) Pritchard, R. H., Kern, C. W.: J. Am. Chem. Soc. *91*, 1631 (1969). — Bonaccorsi,
R., Scrocco, E., Tomasi, J.: J. Chem. Phys. *50*, 2940 (1969).
49) Berthier, G., Praud, L., Serre, J.: Jerusalem Symp. Quant. Chem. Biochem. *2*,
40 (1969).
50) Bonaccorsi, R., Cimiraglia, R., Scrocco, E., Tomasi, J.: to be published.
51) Pack, G. R., Wang, H., Rein, R.: Chem. Phys. Letters *11*, 381 (1972).
52) Riera, A., Meath, W. J.: Mol. Phys. *24*, 1407 (1972).
53) Ardill, R. W. B., Davison, W. D.: Proc. Roy. Soc. (London) *A 304*, 465 (1968).
54) Faisal, F. H. M.: J. Phys. B., Atom. Mol. Phys. *3*, 636 (1970).
55) Gianturco, F. A., Tait, J. H.: Chem. Phys. Letters *12*, 589 (1972).
56) Truhlar, D. G., Van-Catledge, F. A., Dunning, T. H.: J. Chem. Phys. *57*, 4788
(1972).
57) Bonaccorsi, R., Petrongolo, C., Scrocco, E., Tomasi, J.: Theoret. Chim. Acta *20*,
331 (1971).
58) Pimentel, G. C., Mc Clellan, A. L.: The hydrogen bond. San Francisco: Freeman
and Co. 1960.
59) Petrongolo, C., Scrocco, E., Tomasi, J.: unpublished results.
60) Dreyfus, M., Pullman, A.: Theoret. Chim. Acta *19*, 20 (1970).
61) Morokuma, K.: J. Chem. Phys. *55*, 1236 (1971).
62) Morokuma, K., Pedersen, L.: J. Chem. Phys. *48*, 3275 (1968).
63) Kollman, P. A., Allen, L. C.: J. Chem. Phys. *51*, 3286 (1969).
64) Diercksen, G. H. F.: Chem. Phys. Letters *4*, 373 (1969).
65) Del Bene, J., Pople, J. A.: Chem. Phys. Letters *4*, 426 (1969); J. Chem. Phys.
52, 4858 (1970).
66) Hankins, D., Moskowitz, J. W., Stillinger, F. H.: Chem. Phys. Letters *4*, 527
(1969); J. Chem. Phys. *53*, 4544 (1970).
67) Morokuma, K., Winick, J. R.: J. Chem. Phys. *52*, 1301 (1970).
68) Newton, M. D., Ehrenson, S.: J. Am. Chem. Soc. *93*, 4971 (1971).
69) Alagona, G., Pullman, A., Scrocco, E., Tomasi, J.: Intern. J. Peptide Protein
Chem., to be published.
70) Alagona, G., Cimiraglia, R., Scrocco, E., Tomasi, J.: Theoret. Chim. Acta *25*,
103 (1972).

71) Anderson, A. B.: J. Chem. Phys., submitted.
72) Srebrenik, S., Weinstein, H., Pauncz, R.: Chem. Phys. Letters, to be published.
73) Van Duijneveldt, F. B.: Communication at the First International Congress of Quantum Chemistry, Menton, July, 1973.
74) Gelius, U., Roos, B., Siegbahn, P.: Theoret. Chim. Acta *27*, 171 (1972).
75) Almlöf, J., Johansen, H., Roos, B., Wahlgren, U.: USIP Report 72—16 (1972). To be published in: J. Mol. Spectry.
76) Almlöf, J., Henriksson-Enflo, A., Kowalewski, J., Sundbom, M.: private communication.
77) Port, G. N. J., Pullman, A.: FEBS Letters, *31,* 70 (1973).
78) Dreyfus, M.: These 3è Cycle, University of Paris, 1970.
79) Petrongolo, C., Tomasi, J.: Sixth Jerusalem Symposium on Chemical and Biochemical Reactivity, Jerusalem April, 1973.
80) Petrongolo, C., Tomasi, J., Macchia, B., Macchia, F.: J. Med. Chem., submitted.
81) Wernstein, H., Srebrenik, S., Pauncz, R., Maayani, S., Cohen, S., Sokolovsky, M.: In: Sixth Jerusalem Symposium on Chemical and Biochemical Reactivity, Jerusalem April, 1973.
82) Weinstein, H., Maayani, S., Srebrenik, S., Cohen, S., Sokolovsky, M.: Mol. Pharmacol., in press.
83) Maayani, S., Weinstein, H., Cohen, S., Sokolovsky, M.: Proc. Natl. Acad. Sci. U.S., in press.
84) Snyder, L. C., Shulman, R. G., Neuman, D. B.: J. Chem. Phys. *53*, 256 (1970).
85) Bonaccorsi, R., Scrocco, E., Tomasi, J.: In: Sixth Jerusalem Symposium on Chemical and Biochemical Reactivity, Jerusalem, April 1973.
86) Morokuma, K.: In: First International Congress of Quantum Chemistry, Menton, July 1973. Plenary lectures volume, to be published.

Received March 6, 1973

HMO
Hückel Molecular Orbitals

From the reviews:

E. Heilbronner
and P. A. Straub

With 816 pages
DIN A 4
Loose Leaf. 1966
DM 92,—

"In 1961, when Streitwieser wrote Molecular Orbital Theory for Organic Chemists, he drew attention to the very rapid recent growth of interest in this field—seventy papers in the forties, 600 in the fifties, and a corresponding increase in the sixties. These π-electron molecular orbitals are usually represented as linear combinations of atomic orbitals (LCAO) with certain other approximations as introduced by Hückel. The enormous use of these Hückel MOs has now led to no less than three fullscale publications of tables of the relevant coefficients. The present volume, prepared by Prof. Heilbronner and Dr. Straub, is the latest attempt to provide the coefficients in the MOs and certain other dependent quantities in such a form as to be helpful to chemists who have no desire to make these calculations for themselves." (Nature)

Springer-Verlag
Berlin · Heidelberg · New York
München · London · Paris · Sydney · Tokio · Wien

NMR

Basic Principles and Progress
Grundlagen und Fortschritte
Editors:
P. Diehl, E. Fluck, R. Kosfeld

Volume 1
**P. Diehl, C. L. Kethrapal:
NMR Studies of Molecules
Oriented in the Nematic Phase
of Liquid Crystals
R. G. Jones: The Use of
Symmetry in Nuclear
Magnetic Resonance**
53 figures. V, 174 pages. 1969
Cloth DM 39,–; US $ 17.60
ISBN 3-540-04665-8

Volume 2
**H. J. Keller:
NMR-Untersuchungen
an Komplexverbindungen**
22 Abbildungen. III, 88 Seiten. 1970
Geb. DM 32,–; US $ 14.40
ISBN 3-540-04980-0

Volume 3
**O. Kanert, M. Mehring:
Static Quadrupole Effects in
Disordered Cubic Solids
F. Noack: Nuclear Magnetic
Relaxation Spectroscopy**
73 figures. V, 144 pages. 1971
Cloth DM 48,–; US $ 21.60
ISBN 3-540-05392-1

Volume 4
**Natural and Synthetic High
Polymers**
Lectures presented at the Seventh
Colloquium on NMR Spectroscopy.
Held in the Institut für Physikalische
Chemie, April 13–17, 1970, Technische
Hochschule Aachen
202 figures. X, 309 pages. 1971
Cloth DM 64,–; US $ 28.80
ISBN 3-540-05221-6

Volume 5
**R. A. Hoffman, S. Forsén,
B. Gestblom:
Analysis of NMR Spectra**
A Guide for Chemists
63 figures. III, 165 pages. 1971
Cloth DM 64,–; US $ 28.80
ISBN 3-540-05427-8

Volume 6
**P. Diehl, H. Kellerhals,
E. Lustig: Computer
Assistance in the Analysis of
High-Resolution NMR Spectra**
11 figures. III, 96 pages. 1972
Cloth DM 48,–; US $ 21.60
ISBN 3-540-05532-0

Volume 7
**C. W. Hilbers, C. MacLean:
NMR of Molecules Oriented
in Electric Fields
H. Pfeifer: Nuclear Magnetic
Resonance and Relaxation of
Molecules Adsorbed on Solids**
56 figures. V, 153 pages. 1972
Cloth DM 78,–; US $ 35.10
ISBN 3-540-05687-4

Volume 8
**Magnetic Relaxation
Phenomena and Internal
Kinetics of Fluid Systems**
Lectures presented at the Eighth
Colloquium on NMR Spectroscopy.
Held in the Institut für Physikalische
Chemie, April 14–20, 1972, Technische
Hochschule Aachen
Approx. 121 figures.
Approx. 280 pages. 1973
In preparation
ISBN 3-540-05664-5

Prices are subject to change without notice

**Springer-Verlag
Berlin
Heidelberg
New York**
München · London · Paris
Sydney · Tokyo · Wien